In the Shadow of Green Man

My Journey from Poverty and Hunger to Food Security and Hope

In the Shadow of Green Man

ACRES U.S.A.
AUSTIN, TEXAS

IN THE SHADOW OF GREEN MAN
Copyright © 2017, Reginaldo Haslett-Marroquin

All rights reserved. No part of this book may be used or reproduced without written permission except in cases of brief quotations embodied in articles and books.

Although the author and publisher have made every effort to ensure that the information in this book was correct at press time, the author and publisher do not assume and hereby disclaim any liability to any party for any loss, damage, or disruption caused by errors or omissions, whether such errors or omissions result from negligence, accident, or any other cause.

Acres U.S.A.
P.O. Box 301209
Austin, Texas 78703 U.S.A.
512-892-4400 • info@acresusa.com
www.acresusa.com

Printed in the United States of America

Front cover illustration by Ricardo Levins Morales

ISBN (print edition) 978-1-60173-138-8
ISBN (ebook editions) 978-1-60173-139-5

Dedication

This book is dedicated first to my parents, Bernarda Salvatierra and Magdaleno Marroquin, who raised all of their 13 children with devotion and an endless string of stories to last a lifetime. It is further dedicated to my "second" parents, my father and mother in-law William and Jean Haslett, who have been a supporting and true family to me. Finally, this book is for my wife Amy Haslett-Marroquin, and our children William, Nicktae and Lars, who inspired my storytelling.

Reginaldo Haslett-Marroquin

Thank you to Jean Haslett for proofreading and editing this work.

Table of Contents

About the Author   *ix*

Preface   *xi*

Prologue   *xv*

Guatemala's Civil War   *xviii*

1. The Main Street Project   1
2. How Green Man Became Green Man   13
3. Green Man and Mother Blight   27
4. Green Man and His Parents   41
5. Green Man and the Great Snake   59
6. Green Man's Gift   75
7. Green Man and the Garden of the World   85
8. Green Man and the Soldiers   97
9. Green Man and the Strange Fire   109
10. Green Man Among the Ants   125
11. Green Man Goes to Town   139
12. Green Man and the Long Drought   151
13. Green Man's Daughter   175

Index   *187*

About the Author

Reginaldo (Regi) Haslett-Marroquin

Regi is the principal architect of the innovative poultry-centered regenerative agriculture model that is at the heart of Main Street Project's work. As chief strategy officer, his focus is on the development of multi-level strategies for building regenerative food and agriculture systems that deliver social, economic and ecological benefits. He leads Main Street Project's engineering and design work and currently oversees the implementation of restorative blueprints for communities in the United States, Mexico, and Guatemala. He is also leading explorations in Haiti, Colombia, and Africa.

A native Guatemalan, Regi received his agronomy degree from the Central National School of Agriculture, studied at the Universidad de San Carlos in Guatemala, and graduated from Augsburg College in Minneapolis with a major in international business administration and a minor in communications.

Regi's long history with social enterprise development began while working on economic development projects with indigenous Guatemalan communities in 1988. After immigrating to the U.S. in 1992, he served as a consultant for the United Nations Development Program's Bureau for Latin America and as an advisor to the World

Council of Indigenous Peoples. He was a founding member of the Fair Trade Federation in 1994 and supported numerous international social enterprising initiatives.

He served as director of the Fair Trade Program at the Institute for Agriculture and Trade Policy from 1995 to 1998, and led the creation, strategic positioning, start-up and launch of Peace Coffee, a Minnesota-based fair-trade coffee company.

Regi lives in Northfield Minnesota with his wife, Amy, and their children, William, Ana Nicktae, and Lars Decarlo.

The author, center, with his parents.

Preface

I've known Regi for most of my life. I grew up alongside his kids, I helped his wife Amy with the daycare she ran. I ate some of the black beans that Regi grew in their cramped inner-city backyard and watched him toss tortillas on the grill, spinning them with his thickly calloused hands. I heard his stories.

Regi's tales of the Green Man, of life in the forests and mountains of Guatemala, were part of the fabric of my childhood, each story layered and woven together like the many-colored bolts of cloth he sold through a little store in the Center for the Americas building.

When he started Peace Coffee, I learned about Rigoberta Menchu and about her book. I learned about the plight of Guatemala's farmers. When I took up fencing for sport in high school, Regi mentioned offhand that growing up they had only had machetes.

I thought that I knew Regi pretty well right up until the point I sat down with him for the first interviews that became this book. I knew the outline of his life, but this was my first clear-eyed glance into his mind.

It was immediately clear to me, and Regi agreed, that a traditional just-the-facts-style memoir was not the way to tell this story. The circumstances and stakes that he grew up with were too far-flung from the experience of people who grew up in the United States for his story to translate. The realities of life in rural Guatemala followed a different pattern, a different pace. For lack of a better term it is more magical than First World journalism is equipped to deal with.

So I turned to some great Central American writers for inspiration: Isabelle Allende, Jorge Louis Borges and Guillermo Del Toro. They all write stories that are true to their experiences, to their lives, but that seem fantastical to readers in the North. Well, Regi's story is pretty fantastical—a boy who started with nothing making his way through the war, dodging death by a thread, always carrying with him the image of this garden, this life-giving farm.

And so I set out to tell the truth as Regi experienced it, to invest the reader fully in the world that Regi grew up in: a world separated by precious little time and space from the one that I inhabit, but so very different.

The Green Man stories are an integral part of that world. Some of them I took from Regi directly, some I borrowed from Guatemalan folk tradition, and others I wrote myself. They serve to inform and inspire, just as they did when Regi retreated into his fantasies as a child.

Hopefully, this book doesn't just serve to communicate the facts of Regi's life, but to show the power of his dream, and to provide a sort of guide for anyone drifting through the world, an inspiration for another way of life, one rooted in creating and giving. Maybe together we can start a new, more sustainable chapter in our life together on this planet.

Maybe we can all find a little Green Man within ourselves.

— *Per Andreassen*

Prologue

by Amy and Reginaldo Haslett-Marroquin

Green Man stories were born as our eldest son learned to talk and ask for stories. Reginaldo obliged by sharing the stories of his childhood and youth, with the added novelty of a tiny wise creature and the exaggerations that form any good tale.

Over the years, our children have begged their Papa to put the Green Man stories in writing, but they always lost their magic when the words were on the page. That's until Per came along! Per's masterful retelling and expanding Green Man's adventures have finally captured on paper what Regi has been sharing with his children (and a few of their lucky classmates) for years.

The stories, of course, make the most sense within the context of Regi's own life stories. The interweaving of fiction with reality is intentional. This is the way Regi's life evolved. He wanted a different version of an autobiography, and as he said "what would life be worth without imagination in it, not everything that happens in our lives is real, some of our realities live in our minds only." Regi grew up in the shadow of Green Man. His story blends with those stories of the environment that created his hopes and dreams. It is presented as such in this book as stories within a story, jumping from reality to the forest world of Regi's imagination.

Regi was able to turn many challenging circumstances into opportunities to let his mind create another world, while facing a crushing social and political reality. For Regi, Green Man became the creature that he could count on. It was Green Man that would answer the questions which adults would not dare speak out loud ... questions as to why the soldiers would pick up all the young males in the village and returned them in body bags just months later. Sometimes the questions were easier, like what else would grow under the shade of banana and avocado trees, or where the Pacaya palm would produce the best flowers to harvest for food. Green Man knew it all; Regi simply needed to listen, observe and interpret the answers, which were all in front of him, always ready to share themselves, naturally.

Green Man's connection to the natural world around him is the direction we would like to see our real world move toward. A world where people again listen, learn and live with nature, rather than destroy it as if we did not depend on it.

For our "First-World" readers, it would be tempting to read these stories and see only the hardship and struggles of another people. Growing up in extremely poor conditions during a 34-year civil war in Guatemala was a harsh reality. That is not what this book is about. The message of this story is about resilience, having a dream and following that dream, no matter what obstacles cross your path. It's about our human ability to start over and persist, molding and modifying our dreams until they become a reality. But most of all, this book is about hope. Hope for our planet and its future generations. Hope that our children will have dreams that work toward healing this precious home, both Green Man's and our own. Everything we create, build and can touch first lived in our imagination. Let's imagine and build a world that gives life and makes us whole.

Regi currently works on a new Poultry-Centered Regenerative Agriculture System, which is taking shape and growing rapidly under his leadership at Main Street Project (www.mainstreetproject.org). This book is a prologue to a technical publication that is in the works, one where details necessary for people to practice the broadly described system on our website will be more available so more farmers can adopt this way of farming. The poultry system stems from ancient ways of raising poultry, under an original habitat of multi-canopy forests with an abundance of vertical food production systems. The environment is managed to mimic to the extent possible the conditions that resulted from the geo-evolutionary processes that shaped the ecologies of the world and these winged, ground creatures. It is a system where a multitude of foods are harvested and where poultry serves as a critical link in a continual flow of natural energy that enters the system in the form of feed and is then transformed into meat, eggs, fruits, nuts, forests and vegetables.

In the Shadow of Green Man is a spiritual introduction to the life struggles and the background on which the Poultry-Centered Regenerative Agriculture System design is grounded. It was Regi's decision not to publish a technical book as a first step, but to first set the tone and the background that generated the ideas, the persistence, the deep dedication and passion that the system represents.

All of us who worked on this book hope that these stories reach a diverse audience of young and old alike and that the readers take from them an understanding of how we are connected to the earth and others in so many ways.

Guatemala's Civil War

Guatemala went through 36 years of war from 1960-1996. Regi was born just after the war began and lived through it his whole life, as did millions of Guatemalans. A catalyst to the war occurred in 1954 when the U.S. Central Intelligence Agency backed a coup commanded by Colonel Carlos Castillo Armas against the democratically-elected president, Jacobo Arbenz. Arbenz was considered a communist threat after legalizing the communist party and moving to nationalize the plantations of the United Fruit Company. After the coup, Castillo was established as president. He reversed land reforms that benefited poor farmers and removed voting rights for illiterate Guatemalans. The war began six years later when left-wing guerilla groups started battling government military forces. The long conflict was marked by abductions and violence, including mutilations and public dumping of bodies. The war continued to intensify during the next decade as military-backed presidents fueled the conflict, giving the military more control over civilians. During the 1970s a series of military-dominated governments escalated violence against guerilla groups and indigenous communities.

Post 1982, the war became even more bloody after General Efrain Rios Montt seized power following a military coup. He annulled the 1965 constitution, dissolved Congress and suspended political parties. Montt formed local civilian defense patrols alongside the military in the country and rural indigenous regions; thousands upon thousands of indigenous civilians were killed. It wasn't until 1994 under President Ramiro De Leon Carpio, the former human rights ombudsman, peace talks between the government and rebels of the Guatemalan Revolutionary National Unity began and in 1996 under President Alvaro Arzu, peace negotiations were finalized. The decades of violence have left a terrible toll that is still felt today in Guatemala.

This civil war, like so many wars, had to do with the ownership and control of resources and the destruction of structures that seek an independent native-based ideology for this ownership and control. Native philosophy does not align with the foreign and local corporate structures and corresponding government systems that support such structures. For Guatemala, the problem was the democratic election of its president in the early 1950s. Jacobo Arbens did not share the idea that foreigners should have control in Guatemala's resources management and the profits that they generated. Like in every country where this has happened, the oligarchs and those in power who stand to lose their position of ownership sought help from their foreign allies to remove him from power.

The community of Guatemalans who sought to make changes to shift power were up against local infrastructure and the corresponding international infrastructure created to hold on to an old system that creates poverty while degenerating the very resources that could create wealth and well being. Chief among these resources are food and clean water. Many have been confused with the ideological (communism vs. capitalism) propaganda when truly the battle is about power. Guatemala continues to be sacked of its wealth by mining companies that pillage local resources, destroy the ecologies on which communities have depended for centuries, and leave behind desolation, divided communities, conflict, violence, polluted air, undrinkable water, bare lands ... while inflating someone's pockets elsewhere.

Large armies are needed when a people's ability to hope for a better life through hard work is decimated to the breaking point. It is then when they must be subdued, by force. Our inability to envision and build a better world and our extreme capacity to cause destruction are out of balance. To find our center we must fight, but with the more powerful tools that Creation has gifted us—compassion, imagination, creativity, self-reliance, resilience and science that solves real problems—and turn all of these into regenerative ecological and economic management systems.

Chapter 1
The Main Street Project

South of the Twin Cities, Minnesota dissolves into a patchwork of corn and soybean fields carved out by golden gravel roads. Greying farmhouses stand watch with American Gothic eyes and irrigation systems paint rainbows in the air.

Following the highway is a meditation in industrial agriculture, row crops make for trance-inducing scenery, which in no way excuses me for getting lost twice on the way to the Main Street Project's research and development farm.

When I managed to strike the right road, I recognized the farm before I spotted the hand-painted sign. It is wild in a way that even the edgiest organic farms aren't. Plants are densely intercropped in rows, with wild grasses struggling to fill in the gaps. Chickens peck their way across paddocks shaded with hazelnut bushes, and every square inch of land is stamped with imagination.

The other clue that I'm on to something different here is that the corn is two feet taller than the next field over. There's a hedgerow of the biggest asparagus I've ever seen, still producing new shoots in August.

Pulling into the driveway was a strange sort of homecoming for me. This is clearly the farm of the family I grew up next door to in Minneapolis. Their house in town always seemed like a farm crammed onto an inner-city residential lot. The Main Street Project farm is a natural extension of that place, a coming together of the Iowan school teacher and the Guatemalan farmer.

Amy greeted me with a hug, all flaming red hair and smiles and assured me that it was alright I was running late. "You know Regi, he can always find something to do. He's out in the field somewhere."

I thanked her and walked off toward a plastic-sheathed greenhouse. I could hear a rhythmic thumping coming from behind the building, and I could see the beginnings of a foundation trench.

Sure enough, I found Regi with his head down compacting the trench with a length of pipe attached to a heavy steel square. In the trench Reginaldo Haslett-Marroquin looked like a part of the earth, a brown wide-brimmed hat was pulled low on his head, the sleeves of his work shirt were pushed up betraying the sort of quiet strength that isn't built in gyms. I was reluctant to intrude, but as soon as he sensed me, Regi hopped out of the trench to wrap me in a hug. He was a little more grey than I remembered, his laugh lines a little bit deeper, but he was as vital and quick to smile as ever. "You found us!"

"Eventually," I said, "what is this?" I gestured at the trench.

"It will be the foundation of a new chicken coop. Look at this soil," he knelt and pointed to the dark earth. "See how black that is? Full of nutrients, and it's still decaying so you can't build right on it. We'll fill this with sand and pour concrete footings. The building will rest on that mostly." His accent is light, and he has a way of talking that conjures up images in the minds of his audience. I could see the coop taking shape in the wave of his hand. "The chickens are the key to the whole thing. We can talk about that later, why chickens. Let me show you around first."

And we were off. There was barely time for me to grab my notebook.

The Main Street test fields are divided by experiment. The popcorn crop occupies the largest area, but we started in the greenhouse. Rows of water beds filled most of the building, which was set up as an aquaponics system. Vegetables sprouted happily in homemade racks and we had to pick our way carefully over snaking PVC and rubber tubes. Regi spared hardly a glance for the growing plants, stopping only long enough to check that I was clear on the concept of aquaponics, a system in which fish and plants are grown simultaneously. Instead he led me to the back room.

"These are the fish tanks."

The tanks themselves are nothing to write home about, yellowing plastic hooked into the rest of the hydroponics system with PVC pipes and isolated by a water drop so that the fish can't directly circulate and eat the plant roots.

"Hydroponics requires the farmer to inject nutrient broth into the system. That's two inputs—chemicals and work. We plan like we're allergic to work, so we use fish. Aquaponics is actually an ancient farming system. We're just adapting it to modern times." As I peered into the tanks the fish roiled to the surface, one even jumped at me. Regi laughed, "They think you're going to feed them."

He scooped feed pellets from a bucket next to the tanks. "That's it," he said, "a few minutes a day to feed the fish, clean the tanks, and harvesting. There's no weeding, no treatments. The whole system is run off of two pumps. It costs more to heat a room in the winter than to run those pumps."

I had to interject, "I don't know, it certainly doesn't look like you're avoiding work. You could have been using a backhoe to dig out that foundation couldn't you?"

"Don't get me wrong," Regi said, "we work hard. But we set up the system so that the work we're doing is in line with our mission and the concepts of regeneration. This way our work comes back to us as energy and we continue the cycle. This way there's less waste; the feed is the only input other than our labor and that gets processed twice; by the fish and then the vegetables. These are just the experiment fish, tilapia are mostly vegetarian; now that we have a working design we'll change to meat-eating fish, that way they'll eat waste from the chickens. Egg waste, guts from processing, all those nutrients will get reintegrated into the system through the breakdown of fish waste. Follow me." He led me back out into the field, I had to duck a spiderweb spanning the top quarter of the door, Regi didn't seem to notice it but I'm sure he did. Minnesota just doesn't have the kind of spiders that can worry him.

"Look at this," he pointed to the lane next to the corn field, "one hundred and fifty feet. Right now it's just space I have to maintain, I have to mow it and remember, we're allergic to work."

As he spoke I could see the landscape changing before me, intention and execution blurred together as he sketched new structures in the air, "So we dig a trench say three feet wide, four feet deep, we turn the corner and dig on that side too. That's…" he paused to do the math in his head, "twenty-seven hundred cubic feet. You can raise over four thousand perch in that, but let's give them some room to swim. Say two thousand perch with over two cubic feet of space and over fifteen gallons of water each. We dig trenches on the other two sides at a higher elevation, we build in water drops to isolate the fish. We would need one pump. Then we cover the whole thing with mesh to keep birds and cats off; we can cover it with shade cloth in the summer, plastic in fall and spring, we can get nine productive months a year, in Minnesota of all places. Plus the structure keeps the chickens in, so we can just release them into the fields everyday and leave them alone, because…"

"We're allergic to work," I said.

He grinned, "Exactly. Now if we do the math, a whole farming process emerges where the actual currency is energy. Put simply, the chickens fertilize the hazelnuts and the intercropped corn or sunflowers, the byproducts feed the fish, the fish fertilize the hydroponic plants, and the chickens will peck at the plant waste. In all of this, all plants and breeds of chickens we choose we watch over their genetic integrity, so no GMOs (Genetically Modified—or engineered—Organisms) plants or industrial breeds of chickens. Our main input is the chicken feed. That initial energy input is reused over and over again as it cycles through the system. That's the formula, annual crops intercropped with perennials for soil health, animals for fertilizer and to kill weeds, limit disease exposure and to produce a maximum amount of food for minimum effort, minimum inputs."

Maybe it was the cloudless blue sky, or the first taste of autumn cool on the breeze, but I found it hard to stay skeptical in the face of Regi's infectious enthusiasm. It all made logical sense, and it was hard to argue with—his corn was producing like wild.

"We're not against anything," Regi said, as he walked me down the closely planted rows of corn. "It's not about being anti-industry, or anti-technology. Look, here's a row of hazelnuts, and there's another one—that's twenty feet. You want to drive a combine harvester through here? You can. It'll pick up the few weeds as well as the corn and sort it all out; it's what they're designed to do. This isn't about going back to the pre-industrial age—we depend on science and technology to advance after all, and build new systems. Regenerative farming is about looking at the advances we've made and creating a holistic system that takes the best parts and gets rid of the worst. Monocultures and synthetic inputs are killing the earth, and small farmers specifically, so we need to engineer a new agriculture system.

"What we stand for is a system that empowers anyone to feed themselves and earn a living, for a system that regenerates the earth rather than robbing it. To build that sort of system we can't compromise productivity and some of the efficiencies we have achieved over the last four decades, as long as they do not violate the principles of our regenerative design.

"I've got as much density of corn plants here as a farmer would put in a conventional field, and I don't have to till or fertilize—the chicken manure takes care of that. I don't have to weed as the chickens eat up all of the ground cover and on top of that, they control bugs that might attack my corn, not that I am proposing that this is the way to produce the corn we produce today conventionally. The key is to first remove the demand for corn as animal feed by growing larger animals the way they were designed by nature, mostly on pastures, then grow smaller amounts of corn as originally intended and turn it into actual food." As we walk and talk he does pull weeds, out of habit and something to do with his hands more than anything else. "If we were to look at this field and think about the corn only, one can say that I lose some ground because of the chicken coops and the hazelnut rows, but even then, I'm still producing seventy-five percent of the corn the neighbor produces conventionally as a monoculture, and that's compared to a good year. My soil is better, my plants are healthier, but let's say I add my real output from the same space. On top of the corn I've got the hazelnuts, I've got the chickens, and we're experimenting with intercropping elderberries or other productive brush that does not mind the hazelnut shade and have the height to stay off the chickens' vertical range. And my input costs are lower. We're competitive with the industry model, but more importantly, we're growing food efficiently, if measured from the real currency of efficiency, energy in, versus energy out."

At the end of the field there were a few rows of corn in worse shape than the rest, "This was an experiment, most of the field we weeded once when the corn was just starting. These two rows we didn't touch." The stalks were thinner, shorter and yellowed, and flowering weeds sprouted around them. "We did nothing, but feel the cobs—they're fully productive, the kernels are well developed, the tassels are healthy. That's because the soil is so good, the chickens have been building it for years. This row," he pointed to a row of corn as healthy as the rest of the field, "we didn't weed this one either, but what we did was come in and mow it once the corn was almost a foot tall. We mowed it pretty high, and the corn came right back; it came back much faster than the weeds so all it had to fight were the grasses. Much less work than weeding, and just as effective. In poor soils weeds would grow faster and choke the corn, but if the soil is healthy, the corn—a grass itself which has lost its wild genes after being domesticated—will outcompete the wild grass, at least in this place for these rows, that much is true."

"How come that's not just the way things are done," I asked, "if it's so much faster and as effective…"

"People are taught to think about food in certain ways, about farming in certain ways, and it's hard to break out of that mindset, besides the industrial system does not care about pollution, and actually spraying a field to kill weeds and insects is much easier and faster and probably more effective than what we do. The consequences of doing that is where the problem is, but because it's always into the future, the direct consequences of industrial agricultural practices always get turned into a conversation about something else, like feeding the world and cheap food, neither one possible through industrial agriculture in net terms. We try to look at things in a different way. For example, when I was in ag school in Guatemala they taught us that it was bad for corn to grow this tall. The wind would blow it over. But we thought, what if we created a friendly environment for the corn and let it grow, what would it need? Plenty of nutrients for a strong stalk, we don't till the soil so the roots have a firm grip, and the hazelnuts serve as a windbreak. Plants want to produce; we just have to give them the cues that it's safe to do it. It's the same thing with the chickens, we watched them, learned from them, and we've adapted to suit their needs instead of forcing them into an environment that we think is somehow better."

"What did you learn from the chickens?"

"Well they're jungle animals, but they also venture into tall grasses where they can also find cover under bushes like hazelnuts to keep the sun off of them and to hide from hawks. They eat at ground level and they want bugs, tender shoots and fresh sprouts, and to scratch and run, so we manage the ranging paddocks to give them just that and ensure that those crops also give us nutritionally and economically valuable products."

"It seems so obvious."

"It is," Regi shrugged, "but no one is doing this on any kind of scale."

Next to the cornfield is a bare patch of ground scattered over with dry husks of plants, "We had to harvest beans in a hurry," Regi said. "It was raining and if we didn't get to them they would rot on the vine."

"How many beans did you get?"

"Out of three hundred square feet? Two hundred pounds dry, enough to feed three families for a year." Regi bent and pulled up one of the dead plants, the dry pods rattled against each other. "Seeds," he said, answering a silent question. "If you wanted to feed people then you planted black beans and corn, that's all we needed as a base meal when I was growing up. The beans are easy; pick them, dry them, sort them. In the first months you don't even have to pre-soak them, just boil and eat. Corn takes work but you can do so much with it. They both keep forever if you're careful. And we don't use GMOs so we save some of the plants to seed next year. These are descendants of plants I was growing in Minneapolis."

"Those were home grown?" I asked, as memories flooded back to me of hand-tossed corn tortillas wrapped in a towel and served around their yellow-walled dining room, of creamy black beans with a nuance and depth of flavor that spoke to me even as a child. I've tried over the years to recreate those meals but no matter how many hours I spend soaking and cooking the beans, it's just not the same.

"Yes. We all went out and harvested them together, the kids always loved that. The family bringing in the food we'll eat for the next year, it becomes a sort of spiritual experience. Anyway, this plot isn't much to look at now, let's see the garden."

The garden is packed with all kinds of vegetables. Volunteer garlic sprouts in and around tomato vines, and kale grows in the shade of more elderberry bushes heavy with fruit. We stopped and picked cherry tomatoes while Regi talked and I tried not to be distracted by the little explosions of flavor.

"That's about it. This garden goes through cycles of fertilization and weeding by chickens. I'm getting new asparagus shoots in August, and the potatoes we planted fought off an infestation of beetles. We didn't spray, we didn't kill the beetles, we just planted the potatoes after the first hatch of beetles had gone looking for potatoes somewhere else; then those that came back were a bit late to cause the extreme damage that would have killed the potatoes and they were healthy enough to keep growing and producing.

"Of course no system will work without pollinators," he waved his hand at wooden boxes with three active beehives set up on stilts just a little ways away. "We let the chickens under there too, they help keep parasites away."

I popped another tomato in my mouth and thought this was about as close to heaven as I could get.

"We should get moving, they'll send out scouts soon. We don't want to make them nervous," Regi reminded me.

"You pay attention to what the bees want too."

He nodded and we picked our way out of the garden.

"Are you ready for lunch?" Regi said.

"I'm a grazer, so those tomatoes will hold me a while, but I could eat if you're hungry."

"That's a dangerous question to ask me," Regi said, "I'm always hungry. My mom used to joke about it."

I had always known that the world Regi grew up in was very different from the one I knew but it struck me then. I took for granted that there would always be something to eat; six, ten times a day, whenever I wanted it. Regi's body had been trained by need and he has a sort of hunger that I've never felt, a hunger that comes from not knowing when you might be able to eat next.

"How did you come up with all of this?" I asked, spreading my hands to take in the farm.

"It started with a philosophy, part practical and part spiritual. I knew I wanted to feed people, to create a system that will allow poor people to get out of poverty. And I wanted to do it in communion with the earth.

"Too many organic farmers are just growing expensive food for rich people; I didn't want to do that. So we aren't anti-science, we aren't anti-production. The beans I grow, for example, are the product of fifteen years of selective breeding and cultivation so that they adapt to Minnesota's climate, its soil, seasons, biomes, etc. We use hybrid strains of elderberries and hazelnuts because the wild plants don't yield fruit in the quantities that we need, but the strains we use produce viable offspring. They are the product of selective breeding, not genetic engineering that stops or degenerates the foundational eco-/geo-evolutionary blueprint of the species.

"We draw the line at GMOs because they fundamentally interfere with natural evolutionary processes and create dependency designed for control, ownership and convenience. GMOs break the cyclical nature of our system by requiring new inputs each season. Those inputs also degenerate health of the biomes by throwing them off balance and they drain the soil of nutrients instead of regenerating it the way plants will if they are part of a fully functioning ecosystem instead of part of an extractive and exploitive design.

"Spiritually we avoid pesticides and that sort of thing because we want to operate from a position of life rather than one of degeneration and death. A few of my bean plants got attacked by a fungus but they were tough enough to produce anyway, so the fungus doesn't bother me unless it gets to the point where the plant can no longer heal itself. There's a wealth of scientific data out there about how to strengthen and repair plants, starting with cultivating a strong soil environment. Now, if the fungus did progress to a point where the plant was suffering, we would have intervened using an organic spray with a specific bacterium that eats the fungus, and then, when it's out of food, the bacteria die off."

"Like penicillin?" I said.

"Close enough."

"Alright, but where did the philosophy come from? You had a traditional industry education after all…"

"That's a long story," he said.

"Then let's get lunch first."

Over sandwiches at the co-op in town he started slowly, "You have to understand — so much of what I knew about the world came from the stories told around the bonfires at funerals, under our forest shack where we camped out for weeks while tending our fields. The rainforest was home to more than just howler monkeys and tigers; there was El Cadejo, the demon dog with a sparking tail who would take you away if you couldn't learn to live with him, and there was La Llorona who wandered crying, in search of her lost babies. The one that frightened me the most was La Siguanaba who would appear as a beautiful woman and lure men away into the forest until at last, when they were all alone, she would turn around and her face would be the face of a horse."

"You knew they weren't real though?" I said.

"How would I? My parents couldn't read or write, we barely had clothes to wear. All we had were stories. Our parents and cousins, our neighbors all told stories born from their experiences, or passed on from their teachers. We listened and absorbed them, and found our own place in the narrative."

"How did you survive?" I asked.

"We were farmers, and my father knew the rainforest. And sometimes when we were delirious with hunger, when we'd been walking for miles and had miles to go before we rested, when we were surrounded by the steaming jungle but couldn't find fresh water… We told our own stories, and they kept us going, they became our memories real or not, when the mind is pushed like that there is no difference." There's no bitterness in Regi's voice, he's not trying to impress me with the hard hand he had been dealt.

"What kind of stories?" I said.

"Do you want to hear one?"

"If you don't mind."

"Alright. Now, of all the creatures in the rainforest the vulture is the wisest, the jaguar is the deadliest, and the peacemaker is the Green Man…"

As he spoke the walls of the co-op faded away. Green shades deepening, becoming vivid leaves, and jaguars populated the shadows. Regi's face softened the lines of work and laughter and pain all blended back as the story took hold of me. A story of his youth and hardship, and the fantasies that got him through.

Chapter 2

How Green Man Became Green Man

That year there was no dry season. Beans rotted on the vine, and mudslides carried away whole villages. That year children never stopped coughing and their parents could do nothing but pray for the rain to stop.

The forest thrived that year. The leaves had never had more color and if you stood perfectly still you could hear the plants straining to grow, and there was a farmer who saw it and said that it was beautiful.

"Green," he said over coffee at his neighbor's house, "It's all so green."

"Yes," his neighbor said, "The green of mold and copper water. Fever green."

And the farmer could just shake his head, "Thank you for the coffee," he said.

"I don't see how you can be so calm," the neighbor said. "How will you feed your new wife? And with a child on the way. I'm glad to be old, my children are their own trouble now."

"I've got a feeling," the farmer said, "it will all be alright."

"Well, when you figure out how to plant your feelings and eat them, you let me know. I feel a lot of things." That much was true, the farmer's neighbor had a deep love of money, and he was jealous of the farmer's beautiful bride. He kept his hands soft and clean with expensive soaps, and wore finer clothes than anyone in the village.

The farmer, on the other hand, was poor and his hands were rough. He was never happier than when he was working the soil, planting seeds or building something. He wanted to build a circular house, with doors and windows facing every direction so that no one would feel unwelcome in his home.

"Let me make you an offer," said the neighbor. "I have a dry store of beans, but this weather is too hard on my joints. If you will go into the rainforest for me, plant them and harvest them, I'll give you the first fifty pounds."

The farmer set his mug on the table, "I knew it," he said, "Thank you, thank you." And he shook the neighbor's hand with both of his own.

"You'll get started tomorrow?" his neighbor said.

"If the weather is fine," the farmer said, and his neighbor smiled showing teeth cracked and yellow from too much sugar.

When the farmer told his wife of the arrangement, she looked at the floor. "What's wrong?" he said.

"There's more to this offer than meets the eye," his wife said. "When has he ever done anything selflessly?"

"These are extraordinary circumstances," the farmer said. "Surely he knows that if we don't all pull together the village will vanish."

"Still, I don't trust him."

"What choice do we have?" the farmer said. "The rain has destroyed all our seed crops."

"At least take me with you," his wife said and at length the farmer agreed.

The two of them spent the night making tortillas and the morning dawned bright. and clear. The farmer and his wife took the seed beans from their neighbor's barn and carried them into the forest. They walked for miles surrounded by the breath of the forest, the brushing of leaves on leaves.

As they walked the farmer was lost in the color that he saw all around them, and he would leave the path to cut a spray of leaves or a certain flower for his wife, "So rich, so green," he said again and again, "so alive."

But his wife could not shake a sense of foreboding, "I think we're being followed," she said at last. The farmer didn't seem to hear.

It was already late when they made it to the field, and the shelter was slumped and falling in. The farmer began gathering new palm fronds and bundling them for the roof, while his wife built a fire on the ground to dry out the shelter. By the time the sun set they were hanging their hammocks in the relative safety of the shelter.

They stayed in the forest for a week, clearing the field and sowing new seeds. They ate tortillas spread with lard, and at the end of the week they began the long march back to the village. Before they set out, the farmer's wife put a hand on his arm, "We won't be safe in the village," she said.

"We aren't safe here," her husband laughed, pointing to tiger's prints in the earth around their shelter, "All we can do is be careful, keep our eyes open, and sleep off the ground."

His wife smiled, "Still..."

"It will be alright," her husband said.

Regi stumbled and almost fell. It was ten kilometers from his village to his family's fields and the path, such as it was, never seemed the same twice. Hidden roots tangled along the ground and what little daylight filtered to the forest floor only served to deepen the shadows.

Ahead on the path his father turned, "Are you alright?"

"Yes, yes." Regi wouldn't admit it, but the truth was he was exhausted, and he'd retreated into fantasy to escape the gnawing in the pit of his stomach. Back in his skin now, there was no avoiding it.

"It's not far now," his father said.

Regi just nodded and adjusted his load and pressed on. The woven rope straps cut into his shoulders and his back was raw from the rubbing basket. The basket was woven from strips of plastic, scrounged from the market in the regional capital. It held bundles of tortillas tied up in towels, a battered tin cup, kerosene for the torch, a pound of black beans, a water jug, and his hammock. He wore a pair of dirty shorts held up with a rope, and in his right pocket was a slingshot and a few small stones. His left pocket had a hole.

The ground was soft under his calloused feet, black earth rich with decaying plant matter.

"He always says it's not far," Regi's older brother Alfredo said, "From the moment we leave the village it's 'not far now.'"

Regi flashed him a weak grin, "Maybe it wouldn't be so long if you didn't spend the whole walk complaining."

"Don't let the kerosene spill," Alfredo said, "otherwise I'll have to walk back into town tonight and get more tortillas to replace the ones you ruined." There's a hint of pride in his voice, the same pride he showed when he first learned to read and complained about how long and boring all the books that the school owned were.

Regi adjusted his pack and kept walking, he knew the taste of kerosene-soaked tortillas too well to appreciate his brother's joking. Their father was waiting at a bend up ahead. When the brothers caught up he smiled, "Look for yourselves," he said, "I told you it wasn't far." With a sweep of his hand he ushered them past a stand of trees, and into the new growth of last year's field.

It wasn't quite a clearing. Young ferns had moved in, their stalks thick but not yet woody, and saplings were beginning to fight for light. Above them the canopy was thinner than in the forest proper, and enough light made it to the ground so Regi could feel its warmth—not just the oppressive humidity.

Across the field was a shelter, or the remains of one. Five small tree trunks planted upright, a sixth was knocked over, and so the attic of branches and vines slumped dangerously toward the one corner. Most of the roof had been battered away by the rainy season leaving nothing but a few shreds of palm frond to cover the frame.

"Well," Regi's father said, "let's get to work then."

They raised the sixth trunk and repacked the earth around its base. They cut and restrung the vines that held the tabanco, a lofted upper level, in place and Regi's father headed off on his own to cut palms for the roof. They worked with machetes, by the end of the season their hands would be thickly calloused between the thumb and forefinger but now it rubbed raw.

The attic was restrung before Regi's father made it back, and the boys amused themselves by shooting their slingshots at the wild chickens that sometimes wandered into the field. Regi's brother was the better shot, but even he struggled to hit the paranoid birds. After a few shots each they gave up.

"At least we don't have to pluck one," Alfredo said.

Regi's stomach grumbled in response, "There has to be a better way," he said.

Rummaging through their meager supplies, Regi found a wicker basket, he held it in front of him and looked back and forth from it to the chicken pecking its way across the field. "What would Green Man do?" he muttered to himself.

"What?" Alfredo asked.

"Nothing," *Green Man would call the chicken with fish gut and convince it to jump into a cook pot. But I don't have any fish, and I can't speak chicken.* "What are they eating?" He said.

"Who knows? Seeds, grass?"

Regi knelt and started picking through the chaff in the field, new-green weeds and grasses. Beetles and worms slipped through the cover of the loose topsoil. Digging with his fingers Regi broke up clumps of soil looking for tempting grubs. Whenever he found one he threw it in the basket.

"You want them to just walk into the basket for you?" Alfredo teased.

"Under it," Regi said, standing and clapping the dirt from his hands. Across the field the chicken startled into flight for a few feet and then settled down.

"You'll need a stick or something," Alfredo said, "to prop it up."

Regi was already looking. He found a straight twig, and cut a Y in one end. He set the other end lightly on the ground and balanced one edge of the basket on the stick. He set the grubs on the flat stone and set it under the basket.

The two boys watched for a while in silence. The chicken showed no interest.

"It won't work," Regi's brother said, "You'll need a trigger to drop the basket."

"The basket is the trigger, the chicken bumps it and the stick will fall."

"Maybe."

The wind shifted, and the chicken seemed to catch the scent of the vulnerable grubs. Its head úped up and swiveled this way and that, trying to hone in on the food. The boys kept still, Regi barely dared to breathe. The chicken began to make its meandering way toward the basket.

"Come on, come on," Alfredo whispered like an incantation.

When the chicken had nearly found the basket the wind shifted again. A wall of sound crashed over the field, gunfire and shouted orders. The chicken started and took off for the cover of the trees.

"What is that?" Regi said, trying to become one with the ground, searching the trees for soldiers or rebels or both.

"Probably just an exercise," his brother said, staying sitting. "You hear it now and again when the wind is right. They're nowhere near here or we would have heard them already, and the animals would all be hiding. The soldiers aren't from around here, they don't know how to move with the forest."

"That's dumb," Regi said, sitting up.

"It's so that it's easier for them to kill rebels," Alfredo said knowingly.

"They can't kill them if they can't find them…" Regi said.

His brother shrugged, "That's what they said."

"Who's they?"

"People at school, teachers, the officers that come down from the base. Everybody."

Regi tossed a clump of dirt at him, and his brother laughed. Regi joined in.

"We should start a fire," his brother said suddenly, "for when Papa gets back."

"Fine," Regi said, "I wish we could have caught the chicken though."

"It wouldn't have worked without a trigger."

The two boys continued to call back and forth as they cleared the ground under their shelter and began to stack up the firewood. By the time they got the fire started the wind had shifted again and they were surrounded by the sounds of the forest, the howl of monkeys and the constant whisper of wind through the leaves.

When their father returned his face was grim, but he wouldn't tell the boys what had happened or why he had been gone so long. "I'm here now," he said, "and there's work to do."

The three of them set to repairing the roof. They wove together a great quilt of palm fronds and then Regi's father climbed into the attic space and the two brothers passed the roof up to him. Then they tied it fast to the posts, it was thicker than the forest canopy, no sunlight made it into the shelter, and the slick fronds would repel the rain that still fell almost daily.

Regi was no stranger to the rainforest or to the work of building and maintaining the camp, but this was the first time he would spend the full season farming. His parents had insisted that he attend the village school through sixth grade, and until then his father and older brothers had shouldered the majority of the farm work. Now it was his turn to share the load.

The boys watched the crackling fire while Regi's father filled a pan with dried beans and water and set it on to boil. The fire sparked around the bottom of the pan like the tail of El Cadejo.

"Tomorrow we'll start clearing the field north of here," Regi's father said. "If the weather stays good we should be able to start burning by midweek, we might even get to go home early."

Regi nodded but didn't get his hopes up; on the farm the amount of work there was to do always seemed to grow to fit the amount of time allotted to do it in.

"Tell us a story," Alfredo said while they waited for the beans to stew.

"What sort of story?" their father asked.

"A scary one," Regi's brother said.

"Alright," their father began, "now this is a true story, so pay close attention. It all happened when I was a young man, not much older than you two. Now, my uncles were famous hunters, people told stories about their exploits for miles around, and this was my first time getting to go into the rainforest on one of their expeditions.

"We carried dried tortillas and beans for soup to last us a week, and my uncles each had a rifle—this was back when you could still carry weapons—but all I had was my machete. We walked into the forest and made camp a long way from anywhere that I recognized. The sounds of the forest were all strange to me, even the rain sounded different than before.

"When we'd hung our hammocks for the night I begged my uncle to let me shoot the rifle, I wanted so badly to impress him, but he told me I was too small, I would only hurt myself. Maybe on my next trip. But I was determined, so later that night I only pretended to go to sleep, and when the others were snoring I slipped out, and stole the rifle from my uncle's kit.

"I stole off into the forest, it was dark and I was shaking. I gave up on the idea of firing the gun, I just wanted to hold it, to aim it off into the trees… That's when I heard her. The weeping woman. You've never heard such sadness. I followed the sound until I came across her by a spring all dressed in white.

"'Can I help you?' I asked.

"'Will you forgive me?' she said. I was confused and stayed quiet, 'My child, will you forgive me?' she said again.

"I became frightened, 'Yes,' I said, 'Yes, I forgive you.'

"'Come here then,' she said, and I felt myself drawn to her, she wrapped me in her arms and then suddenly she pushed me down, shoved my head into the water. I struggled but she was so strong, my lungs were burning when I heard a shot ring out. She let go of me and I spluttered to the surface. I watched her run out over the water, and then she was gone.

"I apologized to my uncles but they just hugged me. They told me anyone who could survive an encounter with La Llorona could hunt with them any time. But ever since then, every time I go into the rainforest I can hear something, almost inaudible, just on the edge of waking. All around I can hear her weeping, and I know that one day she will come for me."

The boys stared rapt as their father finished talking. He told the story with such sincerity that Regi was pulled in forgetting momentarily that his father had grown up in the arid lowlands, and that he had heard the same story at the funeral of one of the villagers a few months before.

Regi's father laughed and lifted the lid off the pot of beans. "Dinner's ready," he said.

That night the mosquitoes weren't bad, so they strung their hammocks between the posts of their shelter and went to sleep.

The next few days passed in a haze of stinging smoke and blistered palms as the three slashed down brush and small trees, and torched the piles. The work left little time or breath for conversation, and evenings found them too tired to do much more than eat and crawl into their hammocks. Each morning they banked up a fire in the base of their shelter with a layer of green wood to generate even more billows of smoke.

By the time they returned from the fields each day the fires would have died down, and any scorpions or spiders nesting in the roof would have been knocked to the ground by the smoke. It still paid to be careful, stepping barefoot on a scorpion was no better than having one sting you on purpose, but the most dangerous time was midday when the roof was not yet clear.

Once, Regi's father returned to the shelter for water and a scorpion fell onto his hand. Even in its smoke-induced torpor the scorpion lashed out and stung him. With a cry he shook of the scorpion but his hand was already swelling. Cradling it, he worked quickly to clean the wound and pack cold mud around it.

Afterward he said that the first thing he felt was his tongue going numb, but at the time he was quickly too weak to speak. He fell into a fever that kept him confined to his hammock for a full day. He moved sluggishly for a few days after that, as though he was sore all over.

"When you see a scorpion, look at the claws," he said. "If the claws are large, the scorpion is used to getting his way, he intimidates his enemies, and has an easy time gripping his prey. But you're bigger than him, so you've got nothing to fear. Worry about the scorpions with small claws, they don't have it so easy. They have to work harder, their venom has to be much stronger. You may be bigger than them, but they're used to fighting the uphill battle so your size doesn't bother them."

Mostly, though, no one explained the world to Regi. He learned by watching the adults around him—how they moved through the jungle, what frightened them. The grown-ups in his family were too busy with the problems of day-to-day survival to pass on any but the most immediate lessons, while his teachers had all seemed strangely divorced from the world that Regi knew.

Each morning Regi watched his father shake out his sheet and check his boots and imitated him. The rainforest was full of venomous creatures and farmers who weren't careful soon ended up dead.

"The worst thing you can be, after careless, is afraid," the Green Man said when Regi shivered between sleep and waking, "A scorpion isn't a man, it doesn't want to sting you. A snake doesn't know a rebel from a hunter, or a soldier. They just want to eat, breed, and be left alone. Don't get in their way and they won't bother you."

And so Regi held silent conversations with the scorpions that he shook out of his shoes, before killing them he would explain why he had to do it, "If we could just understand each other," he would say, "You would go on your way and I could go on mine."

Sometimes they seemed to understand, and after a moment's hesitation walked away. Regi would wish them well, and yearn to speak to them in a way that they could understand, to apologize for disturbing them, and to find out what secrets they knew. He imagined wild warrens just under the earth, tangles of tunnels and passages and hidden worlds. Of course, he only spoke to the scorpions when no one was looking, otherwise he might get in trouble for letting them go.

By the end of the week the three had cut a large field and burned it clear, ready to plant. Mid-morning Friday, they packed up their few tools and began the long walk back to the village. It felt good to be heading home, and Regi thought he could almost smell fresh tortillas on the comal. He imagined the fire already crackling in the hearth, warming the round clay cooking surface until drops of water danced across it.

When they were nearly to the village, the farmer's wife stopped and pointed off the path, "Look there," she said.

The farmer looked where she was pointing and saw someone sleeping under a tree, "Now he isn't safe," he said and went to wake him. His wife, sensing danger, hid in the brush off the path.

The farmer shook the sleeping man, whom he recognized as a friend from the village, "Wake up," he said. "Did you forget your hammock? Lazy bones…"

His friend woke with a start, "Thank God!" he exclaimed and gripped the farmer's arms. "You must hide, your neighbor meant to steal your wife while you were in the forest and now he has told the village that you stole his beans. His sons are out day and night patrolling the border with machetes."

"Thank you," the farmer said, his voice was the hollow note of the marimba.

"You should run," his friend said. "I believe you're innocent, but no one else will."

And so they ran, the farmer and his wife, back to the farm and then farther, deeper into the rainforest. As they ran, the rains began again and soon grew so strong that the two had to stop and seek shelter.

"You were right," the farmer said. "What do we do now?"

His wife was breathing heavily and holding her stomach, all swollen with child.

"Right now," she said between gasps, "we have to stop and wait."

The farmer watched confused as she lay in the mud, and it wasn't much later that she gave birth to their child, pushing him onto a banana leaf.

"He's green," the farmer said, laughing in spite of the fear and kissing his wife, "He's so green."

But color wasn't the only strange thing about the child. He was born with a full head of hair, black as the night, and he never cried—he just looked around in wonder and laughed with joy at what he saw.

"What should we call him?" the farmer asked.

"We'll call him what he is," his wife replied. And so they named him Green Boy.

The child grew up fast, like a tiger cub. In a week he could climb into the hammock by himself, and by the time he was a month old he could keep up with his parents when they went foraging. Green Boy could spot edible plants from yards away, and his first word was, "Eggs!"

As he continued to grow, his parents could only wonder at him; he knew the names of plants and beasts without being told and could sense which were poisonous and which were good to eat.

One day, when he was just six months old, Green Boy wandered away from his family's camp in the early, early morning while his parents were still asleep. When they woke they were frightened to find him missing, for despite the boy's strangeness they both loved him dearly.

They didn't need to speak, each took a machete and the farmer and his wife set off into the forest in search of their son. Now, the farmer's eyes were sharp from tending to his fields and he spotted a set of footprints that looked like they belonged to his son and began to follow them, but his wife's nose was keener and she could track a beast by smell like the best of the hunters and when she caught her son's scent on the breeze she broke from the trail and started to cut through the brush. The farmer called after his wife but she didn't stop and he returned to the tracks, but he took care to cut a notch into the trees that he passed so that his trail would be easy to follow.

The hours grew long and the shafts of sunlight lancing through the canopy were angled like knife cuts across the forest shadow. The farmer was growing tired, and he no longer looked at the trees as he made his mark—he had eyes only for his son's footprints. So he didn't see the snake until it had wrapped its first coil around his chest.

The farmer cried out as the breath was being crushed from him. A short ways away his wife heard the cry and came running. By the time she found him the snake was all wrapped around his torso and squeezing as hard as it could. The farmer's wife raised her machete to try to free the farmer, but Green Boy's voice broke like thunder from behind her.

"Stop!" he said.

The farmer's wife turned and so, too, did the snake. The farmer was too busy being choked to notice.

"But son…" the farmer's wife started, but Green Boy cut her off.

"Now why would you go and do something like this?" he asked the snake.

The snake hissed a reply.

"I don't care how careless he was, he didn't mean you any harm so it's no reason to crush him. Besides, he's my father." Green Boy's voice was stern but caring.

The snake hissed again and hung its head.

"Okay, I'll tell him," Green Boy said. "I'm certain he won't do it again."

The snake hissed one last time before dropping to the ground and slithering back to its tree.

"Are you alright?" Green Boy asked his father.

The farmer nodded, gasping to catch his breath.

"She said you were waving your machete around without looking and you nearly cut her," Green Boy said, and the farmer admitted it was true. "Well that was a silly thing to do," Green Boy said, "and I'm just glad no one was actually hurt."

His father didn't argue the point.

"Now, follow me. I've been working on something all day."

And with that, Green Boy led his parents through the forest to a small clearing where an underground spring broke through the surface to bubble cool and crystal clear. Next to the spring Green Boy had cleared a large patch of ground and there he dug twelve holes. In the holes he put twelve tree trunks so that they stood in a circle and over the top he had fashioned a roof of palm fronds and branches.

"My house…" the farmer said.

"I built it in a circle so that everyone would feel welcome," Green Boy said. "Do you like it?"

"It's perfect," his mother said.

From that day forward his parents and all the animals in the rainforest called him Green Man. The three of them lived quite happily by the spring until the day Green Boy's parents left him alone, but that's a story for another day.

Chapter 3

Green Man and Mother Blight

Green Man knew all the things that lived in the rainforest—from the largest snakes and hunting cats to the smallest bird and the crawling ant. He knew where to find sweet berries and where to dig for wild root vegetables, he could track a storm by the pricking of his arm hairs, and he could understand the secret speech of birds.

Oftentimes, when there was a dispute among the animals, they would bring it to Green Man. The creatures consulted him not only because he was wise and fair, but because there was no one in the rainforest that was at all like him and so he could hear their cases without prejudice.

So it was that Green Man wasn't surprised when he woke one morning to a spider monkey tugging insistently on his hair.

"Green Man, Green Man come quickly," the monkey said. "My mother is very sick."

"What is she sick with?" Green Man asked.

"I don't know," the monkey cried, his eyes were so wide Green Man was afraid they would fall out of his head.

"Alright," Green Man said, "show me the way." Of course he knew where the monkeys lived, but he hoped his good manners would rub off on the creature.

"Thank you!" spider monkey said, as he jumped and caught a vine and swung up into the trees. Green Man followed him on the ground, running smoothly across the uneven forest ground.

Not an hour later they came to a great tree with many thick branches shooting out from its trunk and so thickly hung with vines that Green Man couldn't see its bark. He jumped and climbed a vine to the lowest ring of branches, and there he met the spider monkey.

"My family lives much farther up," spider monkey said, "but mother was too ill to make it to our branch last night; she's resting on the other side of the tree, follow me."

Green Man followed the spider monkey and wished that his feet were as clever as the monkey's; he could read the tree and find handholds that would have eluded the most experienced climber, but he still had to work harder than the monkey just to keep up.

It took them a long time to circle the tree, but the monkey was patient with Green Man, and that more than anything told him that this was serious. The monkeys usually make jokes and don't often turn down an opportunity to make fun of someone in their trees. Still, eventually they rounded the great trunk—many times the width of a man—and they found spider monkey's mother shivering in a hollow.

At first Green Man didn't see her because she was covered all over in what looked like thick grey moss, but then she opened her eyes and cried out, "Oh you brought him, thank goodness." Her voice was thin and rattled with phlegm.

"You'll be alright," Green Man said, though he didn't feel as sure as he sounded. "What happened?"

"I was looking for food on the forest floor," she said, "big mistake. I found a field all ready to pluck, all planted in rows. So convenient!"

"You shouldn't steal food," Green Man chastised her.

"And now I'm sorry I did!" she said, and Green Man had to hide his surprise—spider monkeys never apologized—"But the farmer was gone, so what was I to do?"

The excuse reassured Green Man a little—that was more like a monkey.

"Well," he said, "what did you do?"

"I went through the field, and I ate what I wanted."

"Was there anything unusual about the field?"

"Not particularly," she said, "only it looked like the farmer had been gone a long time, there was mold on everything. Grey-green fuzzy stuff."

"Did you eat any of that?"

"Ew, no," the spider monkey's mother said, "I picked it off and ate the parts of the corn that were still good."

Green Man leaned down and looked closely at the tree around the monkey. It looked like the moss that covered her was spreading onto the tree. The smell of it burned his nose like nothing else.

"Can you help her?" the spider monkey asked.

"Maybe," Green Man said, "but I have to go find someone first."

"Who do you have to find?"

"An old friend," Green Man said, because he didn't want to frighten the poor monkeys. "One who might be able to help."

The lie was the least of Green Man's troubles as he climbed down the tree. The rainforest was home to more than just animals, spirits lived there, and ghosts, and stranger things as well. Mostly they kept to themselves, but Green Man occasionally had dealings with the Elders, and the one that he feared the most was Mother Blight.

Mother Blight lived in the lowest place in the entire forest, where water pooled and no light shone, in the middle of a swamp from which no human ever returned. There she slept, just under the water, her hair pooling around her, and when she breathed out, water spat into the air and became spores, spreading fungus and rot.

Now, so long as Mother Blight slept peacefully her breathing was slow, and she produced gentle rot—rot that turned waste into soil and regenerated the forest; but when nightmares troubled her, Mother Blight slept fitfully and spat out great plumes of fungal spores. Now she must be having bad dreams indeed, Green Man thought.

Even miles from Mother Blight's home the air was thick with spores. Green Man had to wrap a wet cloth around his face to keep from breathing them in and still his eyes watered and his lungs stung. It was all he could do to keep pressing forward and forward, into the heart of the swamp.

At last he caught sight of Mother Blight's cottage, a leaning stick hut all daubed with mud that clung to a little round hillock in the middle of the swamp. The hut looked weaker than the tufts of grass that sprung up around it, but this low no winds blew and so it remained.

Green Man was about to knock on the door when he felt a breath on the back of his neck.

"I thought I told you all to leave me be!" The words ran and roiled like boiling molasses, wrapped around Green Man's chest and held him fast.

"Mother Blight," Green Man said, "what a surprise!"

"No..." the word stretched like warm rubber. Regi's father stood slump-shouldered and stared over the field. There was row upon row of corn and black beans, newly sprouted, less than knee high. It looked little different from the plants they had left behind two days ago, but sprouted around them was chicken weed. Regi and his brothers stared as well. Alfredo, Anibal and Regi had come with their father to the field to weed. But no one expected a field of weeds.

The plant wasn't native to the rainforest and so none of the animals knew what to make of its blister-shiny stems, its little purple flowers. Regi had heard stories of whole fields taken over by the weed. It grew with abandon, choking out native plants, climbing corn stalks and bringing them crashing down, draining the ground of its nutrients for years to come.

"Well," Regi's father said, his voice firming, "let's get to work."

He fell to his knees and began pulling the weeds. He gripped them close to the base and pulled, getting as much of the root out as possible. The boys joined him. It was back-wrenching work but between the three of them they made good time. Regi tried not to think about what would happen if they lost this harvest—about how his youngest brother's teeth had turned black and fallen out the season that mold destroyed their beans before they could be harvested. He tried not to think about eating nothing but tortillas smeared with lard for days on end. He tried not to think about how, if they lost the corn, they wouldn't even have tortillas to eat.

Darkness came quickly, but Regi's father had to remind the boys to smoke out their shelter—none of them wanted to leave the field. Regi coaxed a fire to life while his brother gathered bundles of the chicken weed to pile onto the flames. The smoke from the burning weeds smelled like milk mixed with honey and left to sour.

Regi's father only stood up when it became too dark to see his hands in front of his face. He came to sit by the fire and silently ate the beans and rice that Regi passed to him. It was hard to be sure in the flickering light, but Regi thought he could see tracks in the dirt smearing his father's face, little valleys left behind rivers of tears.

"It will be alright," Regi said, and the words sounded wrong coming from his mouth.

"Of course," his father said, bowing his head.

"You're all so serious," Anibal said. "We've had weeds before, we'll beat this one."

No one said anything more, and not long afterwards the three hung their hammocks. Regi fell into a dark sleep, he dreamed of heartbeats and deep roots. He saw the whole rainforest as a woven blanket, all brightly dyed strands holding each other in place until; pick, pick. One knot came undone, one strand hung lose, someone started to pull it.

Regi could feel the loose string tugging at his heart, unraveling the world around him. He tried to grab it, to hold on, to stop it running out. But he couldn't. Then there were new threads, tying in like worms, eating the cloth that they were supposed to hold true. He dreamed of truckloads of faceless soldiers marching with sacks on their shoulders instead of rifles, spreading deadly seeds.

When he woke the sun had barely risen, but his father was already in the fields. They worked from dawn until dusk but more chicken weed sprung up in their footsteps. The stuff grew like green fire and in just a few days it was obvious that the crops, the corn especially, was suffering. Its leaves were turning brown and brittle, and the stalks sagged as though they were being beaten down.

At dinner that night Regi's father ate slowly; he seemed to consider each bite with great care and patience. He swallowed as though coming to a decision.

"Tomorrow we go back to town."

"What?" Alfredo said, "We can't…"

"We can't keep fighting this thing by hand," Regi's father said. "We'll lose, and that's it. We'll go to see the agronomist back home. We've got a little bit of money saved, it may be enough to buy a solution." His tone is heavy, defeated, his eyes are focused on something far away and indistinct.

Regi and Anibal began to pack away the camp, "We leave first thing in the morning?"

"Yes, but you can leave most of that here. We'll pack light and travel fast, we can make it back by tomorrow night."

It was a little like a dream, the break in their routine. Regi found himself sliding through the forest. The dirt path turned to gravel by mid-afternoon, and the roofs of the village began to emerge out of the valley—thatched rounds like mushrooms, and rusting corrugated steel. Concrete walls and spirals of barbed wire surrounded the homes of those who could afford them.

The Catholic church and mission house dominated one side of town—its Spanish-style facade had once been white, but was now streaked brown by time and rain, the frescos of the saints worn down until nothing but their empty eyes looked out over the village. In its shadow came the municipality buildings, the police station, and a central park with a basketball court, and beyond that the business sector. Beyond the businesses spread the houses of the relatively comfortable, school teachers from Guatemala City, political appointees, dry-season homes for the local ranchers. Around the outside was scattered the vast mass of the village, the sprawling homes of several hundred farming families, the fire pits and public wells around which they organized their lives.

The agronomist was a graduate of the National School of Agriculture, he taught at the local middle school and worked with the local farmers—for a price. He also owned the town's agrochemical store.

Regi's family lived in an outlying village called Ixobel, on the wrong side of the military airstrip. The mayor was fond of calling it a "nest of rats" which confused Regi, rats were far more likely to gather around the discarded food in the market.

Regi's mother was hanging clothes on the line behind their house when the three arrived; shirts, shorts, a few changes of clothes for the kids who were still in school and very little else. She looked up at the sound of their footsteps and her face fell.

"What's wrong?" she asked.

"Inside," Regi's father said. "You boys wait here, we won't be long."

The brothers didn't speak. The air was full of the hum of insects and the smell of coming rain, and the sky was so clear that the sun seemed to glow from the slopes of the surrounding hills. It was a good day for soccer, or for hunting, the sort of day that Regi had longed to be out in when he was stuck in the classroom.

Now he felt detached, like he was watching his body from a distance. He understood that something different was happening but couldn't put his finger on it. He remembered the way that his parents looked whenever a column of military jeeps rolled through town, heading toward the base or carrying soldiers into the forest. It was a look of tense not-knowing. Now he thought he understood the feeling.

"Mom looked scared," Regi said.

"Of course she did," Alfredo said. "When is it ever good news that we were home early?"

"Planting season," Regi said.

Anibal shrugged, "She's not stupid."

Regi slapped a mosquito that had landed on his arm and didn't answer. Their father emerged from the house a short while later with a canvas bag in his hand.

"Here," he said, handing the boys a few coins, "get yourself a piece of candy and meet me in front of the school in an hour."

The coins felt strange and heavy in Regi's hand, but the promise of candy had set his mouth watering and he smiled. "Thank you!" he cried, before darting off toward the main street.

"Hey, wait up!" Alfredo and Anibal followed at a run.

Regi looked back and caught a glimpse of a genuine smile thawing his father's face, before Alfredo grabbed his arm and lept ahead of him.

The three raced to the broad main street and headed left toward the corner store, an open-walled shack tended by men with cuffed up sleeves who spent most of their days awash in radio static and Coca-Cola.

The boys picked through jars of colorful milk sweets looking for that magic candy that was slightly larger than the others. They paid and then slid back onto the streets, fingers already sticky from their melting treasures.

The candies were smooth and tooth-achingly sweet. The boys ate them in tiny nibbles to make them last as they walked the empty streets of the village. School was in session and the farmers were in the fields and so they saw almost no one as they walked. They heard voices carried on the wind, soldiers drilling in the base down the valley, school children laughing in windowless classrooms. All around they heard the barking of dogs and battle cries of rival roosters.

By the time they wound their way to the schoolhouse, Regi's tongue was thick with sugar and he was looking forward to going back to the rainforest. The ag school turned out the best farmers in the world; surely the agronomist would have something that could beat chicken weed.

When his father met them, he wore a backpack with plastic jugs that fed a tube and a hand sprayer. Regi's brother laughed, "What's all that?"

"The agronomist called it a 'chemical solution,'" Regi's father said, "We spray it at the base of the chicken weed and that's it."

"What do you mean, that's it?"

"He said we wouldn't have to weed or anything, the chemical will take care of it," Regi's father shook his head, "We'll see."

It was dark by the time they made it back to the farm, and faeries swam in front Regi's eyes, darting lights he couldn't attribute to any bug or beast. They slept that night without smoking out the shelter, just covered their hammocks in their sheets. Regi spent the night worried that he had tied the sheet too high and that a scorpion might crawl under the edge. When something landed on the sheet, he flicked it away without a second thought, nothing good could fall from the roof.

Eventually he drifted into the empty pit of sleep.

The next morning Regi's father shouldered the backpack and walked up and down the rows, pumping the sprayer rhythmically, coating the chicken weed. Regi and his brother cooked and did minor chores, but mostly they watched their father. He fell into a rhythm, taking a step, stooping down, spraying a clump of weeds, taking another step.

"How does the chemical know what's chicken weed, and what's corn?" Anibal asked.

"Probably the same way the beetles attack the corn but leave the weeds alone," Regi guessed.

"Yeah, probably." Alfredo let the matter rest, but Regi wasn't satisfied by his own answer. The question buzzed like a mosquito around his ears and wouldn't let him rest.

At the end of a row, Regi's father stood and leaned against a tree to adjust the straps of the backpack. A coil of something long and green slithered from a branch and looped around his arm.

"What's that?" said Regi.

"It could be a vine," Alfredo had already unlimbered his slingshot and loaded a stone. Their father had frozen in place, and that told the boys all they needed to know—there was a snake within easy striking distance.

"Can you kill it?" Regi said.

"If I can see its head..." Alfredo crept nearer. A snake was an unpredictable problem. Unlike the jaguar that left paw prints in the mud around their shelter, the snake may not be hungry—but it also wouldn't be scared off. Loud noises or sudden movements would provoke it to strike rather than flee.

Regi took a machete from its place near the fire; he might have a chance to strike it if his brother missed. *Wishful thinking*, he knew he would have to be faster than the snake to hit it. His only chance would be while it had its fangs in his father. The snake shifted lazily and its scales shimmered green to yellow and then back. It was definitely deadly, too fast to kill with a machete. Regi tightened his grip anyway—maybe it would bite him in the arm instead of striking his father's neck. The world had grown silent, all he could hear was the rushing of blood in his ears.

The snap of the slingshot broke the air like a rock on a still pond and a split second later the snake dropped from the tree. Regi's father stamped on its head, but Regi had already relaxed. Alfredo was still the best shot in the family.

He set the machete aside and ran to his father.

"Okay, okay," his father kept saying. "Let's get back to work. Thank you, let's get back to work. Okay, okay…" he was shaking, and it took him a long time to find his rhythm again. Regi wondered what had gone through his father's head during those moments; just a few months ago, his younger brother had been bitten by the same sort of snake while he slept. Tio Ramon had died in the hospital. When you were bitten this deep in the woods, there wasn't much anyone could do.

When the field was sprayed, they ate a late lunch. The beans tasted sharp and bitter, and Regi's nostrils burned, and the back of his throat. The lingering spray in the air made his head spin and he had to sit more often than not.

"What now?" he asked.

"We wait, we go home. When we come back, the chicken weed will be dead." Regi's father set down his tortilla with a grimace, "One treatment he said, then wait two days." He braced himself and finished the rest of his lunch in two bites, swallowing as quickly as possible.

One question haunted Regi on the trip home, and occupied his brain for the following days—*how?* How did it work? How could a bit of noxious water do more work than three men could do in a week? How could humans exercise such control over something as wild and strange as the rainforest?

If it did work, then what other miracles did the agronomist keep hidden behind his counter? Was there something that could revive the avocado tree near their field? Regi remembered the fruit it used to yield, so soft and sweet you could peel it like a banana.... He lost himself in memory, how creamy and rich the meat of the avocado was, his mouth watered and he felt a wave of hunger building in his gut. He tried not to think about food, but it didn't help. By the time they got home, he was disoriented, his head hung between his shoulders, if his father spoke he couldn't hear it.

By the time they were headed back to the farm a few days later Regi had mostly convinced himself that it couldn't work, that the agronomist had stolen his father's money and that the chicken weed would be thicker than ever. There seemed no way that a chemical could do more than fire, machete, and hard labor.

But of course it had. The corn was green and growing again, and the chicken weed had shriveled to nothing. The rows of crops were littered with the rattling corpses of weeds.

"I need to go back to school," Regi said, the thought forming as the words left his mouth.

"Why?" Anibal asked.

"Because I need to know what happened here," Regi said.

"Death," his father's voice was low, almost inaudible, "death came to this field, and death is all that we'll harvest from it. That chemical I sprayed, it's still here, in the soil, on the corn, in the air. How can we eat food that has grown on death? What will the chemical do to us?" His father realized that the boys were listening and stopped, "It can't kill us," he said. "If it did, how would we pay the agronomist when chicken weed invades next year?"

Regi felt like a river bed at the height of the rainy season, swollen and wild. His father had always said that education was the most important thing. It was why they lived so close to the school and so far from their fields; it was why Regi knew how to read and write, how to measure angles and manipulate numbers.

But his father's logic and intuition had always seemed stronger than what he read in books. It was that intuition that kept them alive when they were stopped by soldiers on the road; it was his knowledge that kept them fed. Regi had never seen his father's logic fail, but here was a crop that he knew was poisonous, but he also knew they would have to eat it.

"Never again," his father said, mostly to himself. "There must be a better way."

Regi wanted to help him, but he needed to know more. He needed to know everything.

"Ah, Green Man," Mother Blight smiled a smile full of crooked black teeth. Her lips were cracked and spotted white and her hair drifted about her head a cloud of kinked black spores. "I didn't realize it was you."

"Who did you think it was?" Green Man said, not sure what to make of her sudden change in tone.

"There are people running around my part of the rainforest, cutting trees, setting fires, spraying poison in the air. They look strange and they smell awful."

Green Man just nodded his head, he was surprised that Mother Blight could smell anything at all.

"Anyway, it's taking everything I have to keep shooing them off of my land. I haven't had a decent night's sleep in weeks," Mother Blight hid her head in her hands.

"I came here to ask for your help," Green Man said, "but maybe you could use mine more. If I can get these men to leave you alone, would you do something for me?"

"Anything," Mother Blight said, "after a nap."

Green Man offered his hand and she shook it, then he turned and went off in search of these strange men.

Mother Blight's swamp was empty of Green Man's usual allies; the monkeys and the birds that shared the news with him didn't come here, even the jaguar and the alligator didn't enter the swamp because there was nothing for them to eat. But Green Man was quick and Green Man was smart, and either the spores had become lighter or he was growing used to them because his throat no longer burned, and he could see long distances again.

So Green Man watched the undergrowth for signs of trampling feet and fire. It didn't take him long to find the men's track; there are the wide ruts left by truck wheels and the rainbow trickle of gasoline pooling on the surface of the swamp—enough for Green Man to follow in the dark.

As he went he kept his ears open for the sounds of boots or voices. He stepped carefully, making no noise and leaving barely a ripple in his wake. When he heard coughing, it shattered the stillness like gunfire and Green Man dropped to the ground. Half crawling, half swimming he made his way in the direction of the sound. He could hear someone else telling the cougher to be quiet, and then some mumbled apology. Then he saw them.

There were four men in total. Three had machetes out and were clearing away brush and trees, while the fourth sat perched on the truck and scanned the swamp. The fourth man had a rifle across his lap and was chewing his lower lip like a piece of tough meat.

Green Man kept perfectly still and watched the men for a long time. They had bags of seed in the truck, and barrels of... something that they sprayed into the water and the air and that seemed to keep the mosquitos away. Green Man was glad of that, even if whatever they were spraying made his skin itch.

After many hours the men clustered back around the truck and began sharing out their evening meal. They cooked on a gas stove, rice and eggs and beans. They drank beer from corked bottles.

As they ate, Green Man crept closer until he stood up just a few feet away.

"Hello," he said.

The fourth man lunged for his gun, but Green Man held up his hands to show he meant no harm. "This is lousy ground you've picked to farm," he said.

"We know," said one of the farmers, "but if we can just get the fungus under control, we should be able to grow rice. And besides, we have nowhere else to go."

"Where did you come from?" Green Man said.

"We came from the east, where it hasn't rained in years. Nothing will grow, and the soldiers take all our animals," the farmer said.

"And our sons," another added.

"We had to flee."

"How did you end up in this place?" Green Man said, "You had to cross many kilometers of rainforest to get here."

"It was an accident," the man with the gun said, sheepishly. "We were chased for a long time, and then we were just trying to get out of the forest—it seems everything is out to get us here. But now we're out of gas and there's nowhere left to go."

"Well," Green Man said, "I'll tell you that there's nothing but sickness here for you. But I can show you a way up into the hills where the land is better and teach you to live there, but you must promise not to take more than is your share."

"We'll promise anything," said the first farmer. "It's either that or starving."

And so Green Man showed the farmers how to clear a field with controlled burns, and which plants grew well, and how to tend them.

In time Mother Blight was able to sleep again, and she freed the monkeys from her fungus, and the fields were no longer contaminated. And Green Man didn't go back to her part of the forest again for a very long time.

Chapter 4

Green Man and His Parents

One morning Green Man woke up and his parents were gone. Their machetes were gone from where they normally rested, against the curved wall of Green Man's circular house. At first this didn't worry Green Man, his parents were industrious people; they farmed and took long walks through the forest to see what was going on in the world.

This day they didn't return for the noon meal, so Green Man ate his tortillas alone. He ate them with eggs and cubes of fried potato, with slices of shimmering-clear onions and salted avocado. Green Man always ate well because he knew where to find wild foods, and the birds helped to keep his garden free of beetles and other pests.

Green Man washed his pan and tin plate. He spent time bent over, pulling weeds from the garden. He walked around the outskirts of his little clearing and called to the birds and the forest creatures who were his friends.

"Have you seen my parents?" he asked the Macaw, but Macaw was too busy preening to notice. So he asked the Guan instead.

The chicken looked up from where she was pecking and blinked at him, "If you can't keep track of your own parents," she said, "you shouldn't expect others to look after them for you."

"How was I to know that they would wander off?" Green Man said.

"My parents were killed by a dog. Their bodies were eaten by vultures. I've had children eaten by hawks, eggs stolen by snakes. The rainforest is a place where creatures go missing. I thought you would have understood that." Guan went back to pecking after bugs in the dirt.

"But not mine," Green Man said, to no one in particular. He told himself that it was still light out, that if he was really nervous he could track his parents down easily enough. Jaguar himself had taught Green Man how to hunt and there was no creature who could hide from him.

Besides, his parents were probably fine. They had lived in the forest longer than Green Man, and while they were slower than him, and their eyes were dimmer, they knew the ways of the forest. So Green Man returned to his house.

He made tortillas and cooked black beans. He sliced bananas and fried them with salt and hot peppers. He hung the husks of corn cobs to dry and treated the kernels with lime to grind into masa. He went into the garden and picked fresh tomatoes and cilantro.

Still, by the time it got dark Green Man's parents weren't back. Green Man didn't eat dinner, but covered the food with a cloth to keep insects out in case his parents came back. Then he hung his hammock and tried to sleep.

When Green Man fell at last into an uneasy slumber he dreamed of laughing chickens and aggressive dogs. He dreamed his parents came across a lake and tried to fish, but only caught snakes. He dreamed that he was lost—a sensation he had never felt—and that the animals would no longer talk to him. It was dark all around and there was thunder across the mountains, the leaves whispered bitter secrets to the wind and he was all alone.

That night it rained like the heavens overflowing. When Green Man woke up, his plates of food had not been touched. His parents still hadn't returned.

Outside the world was transformed. Underground rivers had turned to floodwaters with the rain and welled up through cracks in the earth making small lakes and turbulent, muddy rapids that slashed between the trees and then vanished back underground only to boil up some yards away.

Green Man knew that if he was ever going to see his parents again he would have to go looking for them. So he took a woven pack and filled it with things that were good to eat, and he settled it on his back. Any tracks his parents left had been washed away by the rain, but that didn't worry him. There was always a sign, or the sound of disturbed creatures, or the hope that a friendly creature would have seen them.

So Green Man set out into the forest.

Regi woke before dawn and rolled out of bed. The wooden frame groaned as the woven mat of twine relaxed. The air was damp and chilly, and Regi shivered as he padded across the packed dirt floor. He pushed through the screen onto the back porch and breathed in deep. The smell of cilantro blossomed around him and mingled with the musk of rain-drenched earth, the sharp tang of manure.

Regi rubbed the sleep from his eyes and dropped to his knees. He filled a woven wicker basket with fresh sprigs, and the creases of his hands were soon ground in with dirt and green juice. There was a sizeable garden back here, a tangled growth of vegetables and herbs, but the cilantro was Regi's personal project.

The village stirred to life as he walked, women trundled their wares to the market, children carried feed to lowing animals. Most of the men were out like Regi's father, sleeping by their fields but here and there Regi passed a carpenter, a blacksmith, or a baker stoking fires and gathering the ingredients for the day's work.

When he reached the market, half a dozen women called out to him, "Regi! Boy! Is there cilantro today? Have you brought some for me? Good money, good money."

He shared a few words with each of them as they haggled good naturedly over prices, "How are the kids? Well? Good," and parceled out his precious cilantro for a few centavos here, a few centavos there. He left the market with an empty basket and close to fifty centavos—half a quetzal—jingling in his pocket.

By the time he got home the rest of the family was awake and the house was a cyclone of activity. His mother stood in front of the portrait of the Virgin of Candelaria, the eye of the storm, directing traffic and passing food and clothes to Regi's younger siblings.

She managed to push a warm tortilla and a glass of milk into Regi's hands and he made his way to the back of the house, and he had to dodge his sister Sheny as she ran from the house; he quickly wished her good morning.

"Where's she going in such a hurry?" he asked no one in particular.

"She said she wanted to ride her bike to school this morning," his mother said.

"But I…" Regi began.

"Work it out with her." His mother was already turning away, wrapping up lunches.

Regi hurried to the back where Sheny was unlocking her bike. It was a red thing, too large for her, that she had won in a raffle earlier that year. Regi had attached a board across the handlebars to carry bags, books, or boxes.

He put a hand on the bike as Sheny started to mount it, "I'm sorry," Regi said, "I need to take the bike to market or I'll be late to school."

"I never get to ride it," Sheny protested, "and it's my bike!"

"I know, I know, but the women at the market will be really mad if I don't come back with more cilantro. Maybe they'll find someone else to buy from and then I'd have to go back to working in the fields."

"Oh," Sheny said, not really believing him. "The other kids get to ride their bikes to school."

"I'll tell you what," Regi said, "I'll ride the bike in, but you can ride it home, okay?"

His sister sighed dramatically, "Okay."

"Thank you," Regi said, and he gave her a quick hug before dashing back to the garden. In his head he was counting down the wasted minutes, and figuring how much cilantro he could pick and still make it to school on time.

By the time he made it to the market, the first batch of his cilantro had already been sold and the women were eager for more. Regi made the sales as quickly as he could and shouted, "Thank you!" over his shoulder as he peddled on up toward the school.

He slipped through the gates as the last students vanished out of the courtyard and made it to class just in time. The seventh-grade class was notably smaller than earlier years—by sixth-grade students had learned how to read, write, and use math as proficiently as any farmer could want. Most of the boys Regi's age were working in the fields or apprenticed to a trade. Those students that were left had wealthy families or the sort of grand ambitions that drove them to get out of bed early in the morning and earn a working man's wage by the time the school day started.

The teacher at the head of the class raised his eyebrows at Regi's late entrance, and Regi caught a slight shake of his head as he slid into his desk. It wasn't the first time that Regi had nearly avoided the late bell and it wouldn't be the last. He pulled out his books, *Ah well,* he thought, *There's no help for it, Marcelo René Lopez will just have to learn to deal with it.*

"Why are you here?" Marcelo said, standing as the chapel bells rang the hour. "Don't answer me now, it's not a question to approach lightly. You're here to learn of course, but why? Why are you driven to learn, what answers are you seeking in books? What are your ambitions? Think about this question throughout the day, pay attention in your classes to what inspires you to action. This evening I want you to write a mission statement outlining your answer.

"Now, I trust everyone did the reading for last night? We'll turn to page forty-nine in your text, the chapter is Forms and Function of Government..."

Regi blinked. The night's assignment would be easy for him, he could have answered it then and there. There was only one thing that had taken Regi away from his family's fields, and that was the burning need to attend the National School of Agriculture. The ag school wouldn't accept anyone without a ninth-grade education.

"Reginaldo," Lopez's voice grated across the room, "I asked you a question."

"I'm sorry sir," Regi said, "could you repeat it?"

"Were you not listening?" the teacher said.

"I was distracted, sir."

"Hmm," Lopez said, turning away from him, "perhaps we'll move on then, the rest of the class seems to have been paying attention."

Regi spent the rest of the hour struggling to keep his mind focused on sociology. Phrases like "civic duty" and "the rule of law" kept sliding around the room like oil on water with nothing to pin them to.

He had tried to get through the readings, studying by torchlight until his eyes were swollen with smoke, but he couldn't make the concepts presented in the text seem real. All he could think of was that two months earlier his brother was nearly killed for harvesting pine cones without a license.

The Caribbean pine trees grew on poor savanna soil where the ground was rocky and hard. Not even the most desperate people would try to farm the fallow earth and those who laid claim to the land used it to graze cattle or horses. In August and September they paid the military to patrol the land and catch anyone trying to harvest their pine cones—the Japanese used the dried seeds for something and would import them at a good price.

Regi's older brother Alfredo saw the opportunity to make a few extra centavos and so he ventured into the pine forest early in the morning in search of pine cones. He carried a bag and a long stick, and when he found a good tree he would shimmy up the trunk and knock down the green pine cones. He took great care to move quietly and to keep out of sight behind the brush. When he saw soldiers, he would hide his stick and his sack of pine cones and pretend to be looking for firewood. The soldiers never looked closely enough to see the little spots of blood on his arms and legs where they had been torn up by the tree's rough bark.

The patrol came across him when he was halfway up a tree with no way to run, and nowhere to hide. The patrols were made up of soldiers under the command of officers friendly to the landowners. They were told to treat people without a harvesting permit or identification as rebels. Permits were simple, often handwritten notes from the landowner, but the soldiers would carry lists of licensed harvesters so faking one wasn't an option.

At the base of the tree, the soldiers raised their guns.

"Don't shoot, don't shoot," Alfredo said, raising his hands, gripping the rain slick trunk with his legs.

"Who are you?" The leader of the patrol shouted, "What are you doing?"

"I'm a loyal citizen, just picking pine cones," Alfredo said. "You can search my bag, please don't shoot."

"Drop the bag," the soldier said. Alfredo complied, letting the woven plastic sack fall to the ground.

The leader motioned to one of the soldiers, who hurried forward and kicked the bag open. When pine cones started spilling out he jumped backwards and slipped in the mud, losing his footing and nearly losing his weapon. When Alfredo told the story later he laughed at that part.

"Get down here," the leader said. Alfredo made it to the ground in record time. "Come here."

Alfredo took a tentative step toward the patrol leader, and then another. When he was a few feet away the soldier smiled and hit him in the gut with the butt of his rifle. Alfredo collapsed into the mud.

"He's an Ixobel rat," the leader told the rest of his squad, "but if he's a rebel, he's too dumb to be a threat. Get up!" he ordered.

The world swam around him, and there was a hard knot of pain where the soldier's blow had landed, but Alfredo struggled to his feet.

"You have identification?" the soldier said.

"No," Alfredo said.

"Too young?" The soldier said, "I thought so. Now you're going to turn around," the soldier said, "and we will count to ten. By then you better be gone."

Alfredo nodded, and the soldiers laughed as their leader started counting. Still holding his stomach, he did his best to run. The soldier's voice echoed as he counted, "Uno, dos, tres…"

The words lost any meaning as he concentrated on running, putting one foot in front of the other. He lost feeling in his legs, but he sensed he wasn't moving fast enough. He tried to stab the ground with his feet, poured every ounce of energy into pushing himself farther with each step. "Ocho," Alfredo half turned to look behind him, he was still much too close. The youngest of the soldiers was already turning, raising his gun.

And the earth slid out from under his feet. Alfredo hit the ground hard and heard the first crack of a bullet splitting the air where his body had been. The pain vanished as he half crawled, half slid down a ridge to the cover of the forest. He only stopped when his lungs began burning too badly for him to breathe.

The bell rang and Regi looked up from his desk. Lopez grabbed him by the arm as he followed the rest of the students out of the classroom.

"I'll be particularly interested to see what you write," he said.

Regi nodded, and the teacher released him.

The rest of the day passed in a blur; math, literature, lunch. The only things that stood out were Regi's science class and rehearsal for the play that the school's Dramatic Society was putting on. The play was a dramatization of the legend of Tecun Uman, the last Maya warrior, and they planned to perform at an upcoming festival.

Regi slipped into the role of the hero easily; his time in the fields meant that even if he was smaller than the children from the village, he had a physical strength and confidence that they couldn't match. The tedious work had also given him hours upon hours of time to exercise his imagination. The world of Tecun Uman was tragic, and beautiful in its simplicity. The warrior didn't have to worry about scraping together money for books or passing classes that he didn't care about. All Tecun Uman had to do was give his life for his people, and have his soul borne away in the ruby breast of the quetzal.

When Regi spoke the lines his words rang with conviction and truth.

On his way home from school, after playing the great hero, Regi would stop by the home of Don Juan Guzman who raised cattle for milk and meat. He went straight to the back and started mucking out the stalls.

Sometimes Guzman would come and watch him, "You missed a spot," he might say, or, "They say Domingo from down the road was disappeared the other night."

"Oh?" Regi said, trying not to breathe through his nose.

"Mhm, I had heard he was mixed up in stuff. Still, you never think it will be someone like that. Upstanding, quiet man."

"Hmm," Regi said.

"You're some piece of work, you know that kid?" Guzman said, "Most people muck out a stall for me, they don't take their payments in shit. What do you do with it?"

"I've got to run Don Juan," Regi said, "I'm sorry I couldn't get more done."

What he didn't tell Don Juan was that his "shit" was the key to Regi's cilantro business, and that he carefully folded the manure in with kitchen scraps each night, and let it turn to rich soil. He'd read about the technique from a book in the school's library when he was supposed to be studying for a test in on geometry, and he had tried it out with soil from underneath the chicken coop. When he turned shit into the soil he knew that anything he planted would grow fast and furious, and so he depended on it to give the women at the market what they wanted.

That night Regi's father came late to dinner. "I just received a telegram from your big sister," he said, his face glowing, "Tonita just gave birth to her first baby."

All around the hand-carved table erupted a chorus of questions and congratulations. "I don't know, I don't know," Regi's father said, "All that was in the telegram was that they were both healthy and happy. I'm going to go visit and then I'll be able to tell you more."

Regi's sister lived in the highlands where her husband worked for the rural development bank, helping small farmers with loans and advice. He was a graduate of the agriculture school, and Regi had imagined over and over the conversation that he wanted to have with his brother-in-law, but he couldn't figure out how to ask for help. *Besides,* Regi thought, *Papa wouldn't let me go up there even to ask.* A journey up into the mountains on the other side of the country wasn't one undertaken lightly, and the bulk of the evening was given over to preparing

Regi's father for the trip to meet his new granddaughter. They made tortillas and packed food for the road, and scraped together enough cash to cover the chicken bus fares, which he would have to take from town to town. He stored the folded bills about his person as best he could, with just enough money hidden any one place to convince a highwayman that he had nothing else to steal—he didn't want to be caught without money if a policeman demanded a bribe.

"I'll walk into Poptún and catch a bus to the main highway, and from there I'll try to catch a ride into Guatemala City. I'll send a telegram when I get to her house. It shouldn't take me more than two days, assuming no trucks are blocking the roads." There was a quiet laugh around the table. During the last rainy season a truck carrying corn had gotten stuck in the mud on the gravel road between the town and the highway. In the end they had to take the truck apart piece by piece to clear it out of the way.

By the time life settled into its regular patterns again, Regi was aching and ready to sleep. Before closing his eyes, he wrote one line in the notebook he kept for school. "I will graduate from ag school." He fell asleep on the open pages of the book.

Two days later, his sociology teacher held Regi after class.

"That's quite the goal," he said.

"Yes sir," Regi said.

"You know the agriculture school only accepts top students, right?"

"Yes, sir."

"And right now you are barely passing, let alone excelling."

Regi said nothing.

"Do you want to think about a more realistic goal?" Lopez said.

"No sir."

"No?"

"I want to graduate from ag school, and I'll do whatever it takes to make it happen," Regi said. "I'm going to be an agronomist."

"Are you?" Lopez asked. "Well, I'll give you credit for not setting your sights too low… but the road ahead of you is going to be difficult and full of disappointment. No one will fault you for changing your mind."

"What do you think I should be, sir?" Regi said.

"I think you'd make a fine farmer," his teacher said, and waved him away, "Don't be late for your next class."

As Regi turned to go, Lopez looked up from his desk. "Please," he said, "feel free to prove me wrong."

There was silence as Regi lay dying in the village square, surrounded by Spanish soldiers. The quetzal flew on stage, radiant and emerald, guided by the hands of two puppeteers. It landed on Regi's body, weeping as he spoke the famous words, "It is better to die a free man than to live on your knees."

And so he died a hero's death, and the quetzal rose from him, his breast dyed scarlet with his blood. The crowd held its breath as the bird soared into the air, and released it in a roar of thunderous applause.

The other actors filed on stage to take their bows, and Regi rose last. Looking out he saw people from across the region, barefooted villagers from Ixobel—Regi's neighbors—rubbing shoulders with businessmen and teachers from Poptún. Soldiers hung around the edges of the crowd, drawn to the lights and sounds of the festival but nervous with so many of the locals gathered in one place.

High as he was on the energy of the crowd, Regi thought the soldiers looked lonely. He took an extravagant bow. *Not bad,* he thought, *for a skinny-bones farmer,* and he smiled an exhausted smile.

The festival was alive with sweets and dancing and Regi let himself be carried away by the fantasy of lanterns and the prayer-smoke of incense. His father was staying the rest of the month in the highlands, but his brothers were back from the fields, and his uncles were telling stories around the bonfire. Even Marcelo René Lopez congratulated him on his performance, and for one evening everything was right with the world.

On Monday Regi was called out of class. The whole Dramatic Society was assembled in the principal's office; even their teacher looked nervous. When Regi arrived, the principal stood up from his desk.

He was a grey man, heavy-set, with sad eyes. "I got word from the regional commander," he couldn't figure out what to do with his hands, but kept clasping and unclasping them. "He, uh, very much enjoyed your play, and wanted a chance to meet you all. Um," he cleared his throat, "A truck will be here to pick you up shortly."

Regi felt cold, like he was looking at himself from a great distance. The adults in the room were empty as eggshells, and as brittle. No one spoke until two soldiers arrived to bring them to the base.

The drive wasn't long, and the soldiers tried to break the tension by making jokes, but somehow Regi couldn't hear the words they were saying—he was far away, running through the forests that he watched through the truck window, talking to animals and urging plants to grow.

The truck stopped at the gates of the base, and soldiers peered through the windows. Regi wondered what they expected to find. While he waited, he read the sign above the patrol post. In large letters it read "Brigada Militar, Hogar de los Kaibiles" and underneath was the company slogan.

Si avanzo, sígueme; (If I advance, follow me)
Si me detengo, aprémiame; (If I stop, encourage me)
Si retrocedo, mátame. (If I fall back, kill me)

"Who are these kids?" they asked the soldier who was driving.

"Actors," he said.

"What does the comandante want with a truckload of actors?"

"Who knows?" the driver says, "Orders." He shrugged and the soldier questioning him laughed and waved the truck through.

They drove through a dusty parade ground, tramped down by the heavy boots of soldiers. Cement block barracks squatted around the edges of the field and beyond them, larger buildings. Regi saw the red cross of a clinic, and what he assumed was a mess hall. Gunfire cracked from the other side of base in time to shouted commands, and beyond that tanks fired shells into the side of a low hill. During the military festival the parade grounds were a riot of activity, with parachute jumpers and formation drills, but now everything was eerily peaceful.

The truck pulled up in front of a two-story office block, and the driver killed the engine. "Alright," he said, "everybody out. Move it."

The two soldiers ushered them into the building. It was plain and utilitarian, with concrete floors and metal chairs. What airflow there was came through narrow slits set high on the wall and the heat was stifling. The soldiers brought them to the second floor and knocked on a heavy metal door.

"Come," said a voice from inside.

The driver pulled open the door and stepped aside; the other soldier motioned the students through.

The comandante's office was a different world. It was carpeted, and the walls were paneled with dark wood. The comandante was reading through a stack of papers when they entered. He wore his dress uniform, but the jacket was slung over his chair and he'd pushed up his shirtsleeves.

"Ah, yes," he said, setting the papers aside and standing, "You must be the Dramatic Society. Quite the performance."

"Thank you," the teacher said.

"That will be all," the comandante said, motioning the soldiers to leave, "In fact, I was so impressed by the effect you children had, that I have a proposition to make."

Regi looked at the teacher who was stony, joints locked and face frozen.

"Stories are a huge part of what makes a people, what guides them to action, what gives meaning to their life. We want to help spread stories across the region, and we think you are equipped to do that. We'll give you a bus, and an escort from place to place. You'll be the official Military Base's Dramatic Society. What do you say to that?"

Of course there was only one thing to say to a comandante's proposition. And so Regi spent the remainder of the school year studying on bumpy bus rides to different village festivals and celebrations.

The scripts they used changed in more than a few ways, some subtle and some blatant. While still masquerading as a story of Mayan resistance, the script changed to fighting for a strong capitalist Guatemala against a scheming, evil force that intended to turn the country over to the Russians. It became a play about heroic soldiers, not noble sacrifices, and ended with a blatant call-to-arms. When the Tecun Human story was retold, the chorus rising up around the quetzal bird told the audience to keep fighting for Guatemala, to resist the threat of outsiders who would subvert their way of life.

Regi couldn't help but notice that whatever meaning had drawn them to these stories in the first place was lost. They felt wrong. The stories had become manipulative, intended as a threat against those who opposed the military. The audience was encouraged to turn on their neighbors, their own family members, and report them for being disloyal. They were no longer playing for the people who thronged the village square, they were playing for something alien, something other. Audiences noticed it too, and they never got the thunderous response that they received at that first production.

It was while Regi was touring with the production that his father returned, the family feasted—roast chicken, eggs and rice, piled high plates of tortillas, avocado and tomatoes. The whole family was there—Regi's older brothers came back from the fields, his aunts and uncles and grandparents were all arrayed around the table. Work was set aside early and Regi took turns with his siblings carrying water from the well for washing.

Regi's father was tired, dirty from the road, and smiling freely. "Tonita looks good," he said of his daughter. "She's put on weight, you can see it in her arms, her face. And the child… ah. That child will never want for anything. They have diapers…plenty of food. They eat meat every day."

"So what did I miss?"

And so they went around the table with news of the weather, the crops, Regi's brother told him about the triathlon that he won over in Poptún against boys with fancy running shoes. Regi updated his father about his progress in school.

"I'm passing my classes," he said. "I'm doing well in literature and my science classes; I'm even pulling ahead in sociology. At this rate I'll be able to test for ag school next year."

His father's smile faded, "We'll talk about that later."

"They don't administer the test in Poptún," Regi's father said as they sat overlooking the garden. The sun was a thin memory on the horizon and the landscape was a suggestion in purple shadow.

"I know, I would have to travel. I thought I could stay in the highlands with…"

"And what about us? What are we supposed to do while you are off studying, testing, what if you get in and you're gone for years, hmm?"

"You've got my brothers," Regi said. "There are plenty of hands to work the farm; it's not like you'll be needing me."

"All my life I've sacrificed and struggled for you kids, and all I want is for you to succeed. That's why I worked twice as hard so that you could go to school, so you could learn to read and write. But you're educated now, no one can take advantage of you through your ignorance. It's time to join me in the fields, help your younger siblings get off to a good start. Maybe once your sisters are all married you can go to the university."

"I don't want to go to the university," Regi said. "What will they teach me? Politics and useless theories. I am going to ag school and I won't let you stop me."

"What are you saying?" Regi's father's voice was even, but his jaw was tight.

"I'll graduate the seventh grade at the end of the season. I can move to the highlands then."

"You will not."

"I'll pay my way up there, I've saved enough money..."

"From selling plants grown in my garden while you stayed under my roof. You will not go and live with your sister. I absolutely forbid it."

"But you said yourself how well off they are, it wouldn't be any trouble for them. It makes sense, and it will prepare me to do," Regi lost his words, "big, huge things."

"No other man will be responsible for my children," Regi's father said, as though that ended the discussion. And Regi took it as his cue to shut up.

"I'm glad you're back," Regi said.

The next time the Dramatic Society traveled, Regi slipped off to the telegraph office and sent a message to Tonita. Making the arrangements for his travel at the end of the season would have been agonizing if Regi hadn't had his schoolwork to keep him occupied. As it was, it was exciting to slip away and make the arrangements in secret. He felt like a character from one of the imported comic books he saw in the corner store.

His plan was simple: go to the mountains, study intensely in the weeks leading up to the test, take the exam and pass it, get accepted to ag school, give his father the good news and ask forgiveness for disobeying him. The only thing Regi wasn't certain about was whether or not his father would forgive him right away.

At the end of the school year his father was in the field for the harvest. Regi told his mother and his sister Sheny what he planned to do and they both nodded. "I'll pray for you," is all his mother managed to choke out—but she made sure he had plenty of food for the road.

Sheny took Regi's bag and walked down the road ahead of him, so that when it was time for Regi to leave it looked like he was just going into town. As he turned his back on his home he felt torn. He hated to think that he was disrespecting his father, but he could no longer deny the urge that consumed him, the need to go further and be bigger than he was.

"They'll find out where you've gone," Sheny said when Regi caught up with her.

"Kids disappear all the time," Regi said. "Say you saw soldiers taking me away if anyone asks." Normally they didn't, normally everyone knew well enough to mourn for their lost friends, sons, and brothers in silence. The worst thing that could happen was for word to get around that the boy had actually gone to join the rebels, then there would be trouble for the whole family. "When I come back," Regi said, "they'll understand."

"You're not coming back," his sister said through her tears, "you're too stubborn."

Sheny gave him a hug, handed over the bag and walked back to the house.

"I will miss you," Regi said, and he walked into the world.

Green Man searched high and low for any sign of his parents, but there was no trail in the underbrush, and when he asked the monkeys for help they just laughed at him.

"Look at you," they said, "a grown man searching for his parents."

"They are all I have in the world," Green Man said.

"He must be blind," the monkeys said, laughing. "Can't even see his own house or hammock. 'All I have…' silly man."

So Green Man walked on. He kept his nose to the wind and his ears open; he tried to extend his senses through sheer force of will, but he couldn't catch a sign of his parents. He came across a ruined temple, and climbed it until he could see a kilometer through the trees, but he couldn't see his parents.

When the sun began to set Green Man grew hungry, and so he turned to foraging. He found wild carrots, onions, tender shoots and sweet potatoes, and making a fire, he roasted them until they were soft. Green Man burned his fingers on the hot skins of the potato, but he hardly noticed.

When it was fully dark, he wove a hammock out of vines and strung it between two trees. He was so miserable that even the mosquitos took pity on him and didn't disturb his sleep. Green Man tossed and turned all night and he didn't dream.

In the morning Green Man was thirsty and he went in search of clean water. He could drink from streams like the creatures of the forest and not get sick, but he didn't like the taste of mud or silt and so it took him some time to find a river clean enough to drink from. To slake his thirst in the meantime, he drank the water that pooled in the cup-like leaves of the low-lying manaque plants.

The river had burst to the surface with the rain, and so its banks were crumbled and Green Man only walked on tree roots to keep from falling in. Keeping his eyes peeled for snakes he knelt and cupped his hands and filled them with water. The river was cool and tangy and when Green Man had drunk his fill he washed his face and felt refreshed.

Green Man looked across the water and saw something floating on the surface. Curious, he broke a small branch from the tree and used it to draw the thing closer to him. When he saw what it was, Green Man felt his heart break. It was a canvas machete sheath with a belt loop, and he recognized it as the one his mother wore because she had often repaired it with bright thread.

Green Man stood and was about to dive into the river, all heedless of snakes and other unseen creatures, when a voice called out from somewhere above him.

"Wait," Quetzal said, as she came and landed on a branch, "I have seen your parents."

"Where?" Green Man said, "When?"

"Just this morning," Quetzal replied, "but as for where… Your parents were heartbroken to leave you, but they kept reminding each other that it was time. You no longer needed them to protect you, and so all they were doing was holding you back. It's the hardest thing for a parent to do, to set their children free, but it must be done. No one can grow and thrive in captivity."

"But I miss them," Green Man said.

"That's a good thing," Quetzal said, "and so long as you do, so long as you miss them and keep them in your heart then they will never truly leave you."

"Will I ever see them again?" Green Man asked.

"Maybe," Quetzal said, "but that is as much as can be said for anyone. Live your life, Green Man, and let your parents live theirs. Be joyful when your lives cross again, but don't try to force it. Love them from wherever you are, and know that they are loving you."

And so Green Man went home without his parents. But he kept his mother's sheath just in case she ever came back for it.

Chapter 5
Green Man and the Great Snake

That morning Green Man woke up and found that he couldn't move. He couldn't even open his eyes. There was pressure pinning his arms to his sides, and something heavy settled on his chest; it was so heavy that he struggled to breathe.

Green Man was used to running freely through the forest, reading the wind in the leaves the way schoolteachers read books, following the secret blueprints of nature. He tried to move his arms, but he couldn't. He tried to move his legs, but he couldn't.

Anyone else would have been frightened, but Green Man had never met a challenge that he couldn't overcome and so he thought that this would be no different. He took a few deep breaths to settle himself and cast his mind back to the previous night.

He remembered eating dinner on the porch of his round house, and watching the sun drop behind the trees. Dinner had been a chicken that he roasted whole over his camp fire and he took his time picking the last of the meat from the bones. The chicken had lived a long life, and eaten well, so her meat was rich and wholesome. Green Man had left the guts on a smooth rock, and as he ate Vulture began to circle overhead, eventually landing.

"Thank you, Green Man," Vulture said, "I've flown a long way with nothing to eat."

"Were you in a hurry?" Green Man said, "I'm glad you took the time to stop and eat with me."

"No," Vulture said. "It's the strangest thing; it is as though all along my route the animals have either run away or been consumed completely. There wasn't a single natural corpse for me to eat from for more kilometers than I care to remember." And with that, Vulture buried her head in the pile of guts, and for a while she and Green Man ate in silence.

"Do you know what could have caused such a thing?" Green Man said, as the shadows blended into twilight.

"No beast is so greedy that they won't leave something for me. Even humans, who don't know any better as a rule, leave something behind, even if it is out of wastefulness and laziness." Vulture shook her head, "Snakes eat the whole of a thing, but then they lay about until it's gone so they don't eat much."

"Could it be there's some creature that you've never heard of?" Green Man said, "Or perhaps a great many small creatures?"

"Green Man," Vulture said, "everything eats, and because everything eats everything dies. There is no animal in these wide forests that I haven't shared a meal with. And there is no creature that won't pass a word with me. I could as easily ask you if there was a river you couldn't follow or a tree you couldn't climb."

Green Man nodded and picked at the chicken. He was suddenly no longer hungry, but to honor the bird he couldn't bear to throw its meat away. He buried his fingers in the earth and let his senses flow out through the forest like roots through soil, like pollen on the wind, but he could find no answer to the riddle. The weather was fair and the plants sang sweetly to the night. But in a line, for many, many kilometers there was a rotten core of silence where there should have been animal voices.

"I would leave, Green Man," Vulture said, "if I were you."

Green Man shrugged, "If I leave how will I find out what is coming?"

"There are some things that it's better not to know," Vulture said. "Thank you for the meal. I hope we'll see each other again."

"I look forward to it," Green Man said, and watched as his friend spiraled away into the sky. Then he cleaned up and hung his hammock.

Then he was here.

With great patience Green Man flexed all the little muscles in his body, and eventually found that he could move his fingers and toes. Carefully he turned his wrist, and stretched his fingers as far as they could go. He felt something slick wrapped around him, scales over muscles like ancient roots. The coil shifted at his touch and Green Man felt himself being squeezed even more tightly, but Green Man's skin was tougher than tree bark, and his strength was that of a rushing river.

"You are a sssstubborn one." The voice came from far away, "What are you, little thing?"

"I am Green Man," he replied, "and who are you?"

"I am Kukulkan, the servant of the sun, come to restore the ancient order of things."

"And that means eating everything?" Green Man said.

"Yesssss," the serpent replied.

"Ah," Green Man said. He thought for a long time about what to say or do, "Why?"

And so the serpent told him.

Regi set down his pencil and rubbed his weary eyes. Visions swam around him, his tongue was thick for lack of water. He didn't know how long he had spent staring at the problem in front of him.

Tonita and his brother-in-law, Pedro had been more than welcoming. They gave Regi plenty of food, pencils, paper. Pedro even helped him track down materials to study for the ag school test and introduced him to his younger brother Vicente.

Vincente was far ahead of his class in math and helped Regi get up to speed, but mostly he was full of questions. He was fascinated by the idea that anyone actually lived in Peten, a place that Vincente considered a land of monkeys and rebels. Regi, in turn, was envious of the Quiche language that his brother-in-law spoke at home, and the long traditions of the region. The people of the highlands were almost all indigenous. The proud descendants of Tecun Human and other great Mayan warriors. They were still a target of the army—the proud never-conquered people.

In the week before the test Regi tried to block out any distractions, and never pulled his nose out of his books, trying to cram every bit of information he could into his head.

But when the day of the test came he was blank. Occasional concepts seemed familiar but everything was jumbled in his head. The words seemed to rearrange themselves, dreamlike on the page. The test halls were long and stiflingly hot with little air circulation. The other students were rude—to them this was more than just a test, it was a competition and they did everything they could to win. They arrived in their finest clothes, flanked by parents who were alumni of the agriculture school; some even drove their own cars. They all acted like they had already been accepted, while Regi left the school knowing he had failed. It was December of 1983, and Regi wondered if his life might already be over.

"I'm sure it wasn't that bad," Tonita tried to reassure him, but Regi was in no mood to be comforted, so she busied herself making coffee on the stove.

The stove, that was a wonder. It had a small opening to place a few sticks of wood and a chimney so smoke didn't fill the house the way it did back at home. The fire heated a whole steel cover with holes cut into it, grates of various sizes hung above the stove so that you could heat any size pot without worrying that it might tip into the fire and ruin valuable food. She cooked fried eggs with onions and fresh tomato; the smell alone would have been enough to bring together the whole neighborhood back home. She fed the fire generously with wood that other people had chopped and hauled for her. That, Regi thought, is wealth.

"I didn't pass, and if I did… I didn't do well enough to actually be accepted."

"What's next then?" his sister said.

"I don't know. I'll take the test next year, I'll pass it then, but…"

"I'm sure our parents will be happy to have you for another year," Tonita said.

"Hmm," Regi paused, "I can't study there, in the village. There's too much, too much." He opened his hands to the heavens.

"You could stay here…" his sister said slowly.

"Not for free, not stay and do nothing."

"Of course, of course. But Vicente owns a space that is rented by Don Gerónimo who runs a carpentry shop there. Do you think you could work for him?" Tonita set a mug of coffee in front of him, "Take some time to think about it."

But Regi grabbed onto the chance, "I would love to learn. I mean, please, could you ask if he would take me on?"

"We've already talked about it," his sister smiled sheepishly. "Vicente likes you and he asked Gerónimo to let you try and see if you can catch up with the other workers." The baby on her back started crying, Tonita shifted the wrap around her body. "There, there…" she said handing the baby to Regi, who quieted her immediately "she wants you to stay as well."

He should sleep, the sun had set hours ago, and he was studying by the wan light of the carpentry shop's lamp. There was a stack of doors and windows that needed finishing, and he would have to start early if he was to get through them before classes started.

With a hiss of frustration, Regi put his pencil in the book to mark his place and then closed it. He'd boxed his nightly rounds with exhaustion and the angels at the edge of his vision and again he found himself stretched out on the floor, on a mattress of wood shavings. He used a book for a pillow and dreamed of somewhere far away.

Regi got up as the first roosters started crowing. He was used to keeping farmer's hours and welcoming the rising sun with wide open eyes. By the time Vicente left for work in the next town over, Regi was hard at work planing boards by hand and cutting them to length. He worked slowly, carefully, always mindful of Don Geronimo's mantra, "We can always cut boards shorter, but we can't stretch them out."

When the rest of the carpenters arrived at eight in the morning, Regi had already put in almost three hours of work and had laid out the boards that he needed for the day's windows and doors. Putting in more time was the only way that he could keep up with the more experienced men that he worked with.

His sister brought him coffee and breakfast of tortillas, eggs and vegetables, as well as chicken tamales tied in a cloth for his lunch. "Eat," she encouraged him, "you still look like a starving farm boy. If we don't put some fat on those arms, the school will never take you."

So Regi ate. He ate as much as was set before him and anything he could scavenge on the walk from the shop to the night school and back. Most evenings he didn't sleep in his sister's home; he preferred the silence of the shop for studying, and it let him push further into the night without worrying that he would oversleep and be late to work the next day.

Throughout the day Geronimo showed him how to accurately cut angles and to join boards together for the door and window inserts. He learned to use the various tools of the carpenter shop and his hands began to ache from the constant, repetitive work. The windows and doors that the shop turned out were all the same, all meant to be shipped off to new housing projects throughout the rural areas.

Regi was determined never to let his fatigue show. He knew that young men were expected to be tough, to show strength and inspire even when suffering themselves, and so he never complained, and never asked for less work than the men. At some level he understood that no one succeeded alone, that Geronimo's workshop was so busy because of his friendship with Vicente, but he was determined to earn the breaks that he got. The men that he worked with saw Regi's fire, and respected him for it.

The days blurred together sometimes. Regi worked all day and then rushed to class at six in the evening. He would leave the private college at eleven and study in the woodshop until he couldn't keep his eyes open. But there were other things to do that broke up the routine; he spent as much time as possible with his sister, whether that meant helping to care for the baby or carrying feed to the animals. Sometimes he had enough time to take a break in the afternoon and share a cup of cafe con leche with her. When they were together the world seemed to slow down a little, and no matter how long he had been awake he left her company feeling refreshed and energized.

Regi gradually stuffed his mind full of dates, names and scientific principles. He struggled to memorize everything he was taught, determined not to let himself flounder like last time. Vicente was determined to help, and he brought Regi copies of papers on agriculture, logical reasoning, and the like—things that other students took for granted, but that Regi had never had access to. Regi committed two books completely to memory, the two most important books, the books that no applicant could miss anything about: arithmetic and algebra. He often fell asleep reading from one or the other, only to wake up to the sound of the corn mill next door coming to life.

The days blended into weeks, months, and soon the rains had come and gone and the dry season had returned to the highlands. Regi found himself counting down the days until he could take the admission exam again.

A few weeks before the test, Tonita put her foot down, "All the studying in the world won't do you any good," she said, " if you sleepwalk through the exam." She made him stop working at the carpentry shop and simply focus on himself. Regi meditated and went running every day, he relearned how to relax and sleep through the night. He slept on a cot every night, and ate wholesome meals with his sister and brother-in-law.

The day of the test Regi stayed up and took the one a.m. bus to Guatemala City. From there he took a bus headed south toward Villa Nueva and got off at an intersection called "The Tunnel." The school was nowhere in sight. Regi asked for directions from anyone who stopped long enough to listen, and eventually they pointed him in the right direction. He made the three-kilometer hike just in time for the exam, sliding into his seat at eight in the morning.

After the written exam, the aspirants were led into the field where they were assigned rows of corn to pick by hand and load into trucks. Some of the other students groaned as they rolled up their sleeves, but Regi felt himself relax. This was work that was as natural to him as breathing.

The weather was fine and Regi worked quickly—his hands acted with a mind of their own—picking through the plants, finding the easiest place to pluck the ears of corn; his movements were perfectly economical so as to not waste time or energy. Regi could hear the sounds of someone struggling in the next row, and he poked his head between stalks of corn to see what was going on. "Hello," Regi said, "what is your name?" "I am Luis," the boy said.

Luis was significantly taller than him, skinny as well, but his face was flushed and his hands were bleeding. "Are you alright?" Regi asked.

"It's no use," Luis said, "I can't do it."

"Sure you can," Regi said, and he stepped between the rows. "Here, you're trying to yank the ears off. It's easier to grip here and twist." The ear popped off in his hand, Luis looked at him like he had just done magic.

"You've done this before," he said.

"All my life," Regi said.

"I'm from Guatemala City," Luis said, "I've never…"

"I'll finish my row," Regi said, "and come back and help you." He whistled while he worked, he felt confident about the written test, and the proctors had nothing but praise for his performance in the field. As to his encounter with Luis, little did either of them know that this was the start of a lifelong friendship.

The wait for a response was agonizing, but at long last he received the coveted telegram telling him to report to the school in Barcenas Villa Nueva on January 15, 1985. He was to bring with him a mattress and bedding, work shirts and jeans, one pair of work boots, notebooks and pencils.

The following days were a flurry of activity. Regi's brother-in-law helped him find what he needed, used and cheap or free, but the work boots proved to be a problem. For days Regi searched the town for a pair of boots that he could afford, and that wouldn't fall apart after one day in a muddy field.

At last, just before he left, Vicente pulled him aside.

"I have a friend," he said, "who is a soldier. I saw him at the base in Xela, and he sent me these." He took a box from behind his back, "I want you to have them."

Regi opened the box, and inside was a pair of heavy-soled boots, military black with thick straps. They had been worn, but only a few times. Regi held them like they were made of glass; they may have been the most expensive things he had ever owned.

"Put them on," Vicente said. Regi did, he felt powerful and wondered why the soldiers he had seen always looked afraid. "You look like a man," Vicente said.

When Regi boarded the bus the next morning, he carried the boots in his duffel bag, along with the clothes and everything else. He carried the mattress on his shoulder and clenched his teeth to keep from shivering. January was the coldest month and the bucket that he'd bathed with that morning had been covered in a thin sheet of ice. He had washed himself in the backyard and got inside as quickly as possible. Dry and dressed he slowly began to warm up, but he was still stiff with chill by the time he got to the bus.

The bus was parked on the outbound road next to the Catholic church. Regi had his mattress and bag lifted onto the top, where men blew steam from their nostrils and chickens squawked and stalked around a maze of luggage. He took his seat and rubbed his arms to stay warm as the bus gradually filled around him. When at last the bus began moving, a few young men jumped on the back bumper, clinging on for dear life and a free ride.

Regi concentrated on the school. He felt a surreal sense of accomplishment, even though he knew that nothing he had gone through would have been worth it if he couldn't make it through the next three years. His brother-in-law had warned him that the first trimester was the worst. The older students always bullied the newcomers, trying to pressure them into quitting before the first exam. The exam, given three months into the first year, was called "the thinning" and everyone he talked to had told him that was the hardest part—after that it was just a matter of surviving.

The countryside rolled by his window, a pit had opened in his stomach; he was more afraid now than when he had left his parents house that morning that seems eons past. Eventually his feet warmed up again, people around him started to snooze, and the motion of the bus lulled him into an uncomfortable sleep.

He was jolted awake as the bus came to a neck-snapping stop. He was sore from sleeping sitting up, and it took him a minute to realize what was going on. They had hit a checkpoint. Soldiers surrounded the bus, rifles at the ready. The sun was just beginning to rise over the mountains and it shone red through the smoke wafting up from the Pacaya volcano. The soldiers stood in the frost-bound fields with scarves around their necks, gloves and thick green sweaters with green shoulder pads.

"Everyone off," came the order from the front door.

Enough of the passengers had been through this drill before that Regi was able to follow them as they lined up against the bus with their hands pressed against its freezing metal side. He kept his head down as soldiers climbed to the top of the bus and went through each bag, throwing some luggage down so that their fellows could search them more thoroughly.

"Whose bag is this?" One of the soldiers called as he walked along the line of people. Regi risked a glance, and saw that the soldier was holding his duffel.

"Mine," he said without thinking. He turned around just in time for a soldier to grab him and haul him out of line.

"Are you a soldier?" the one holding his duffel barked.

"No sir."

"Deserter?"

"No sir," Regi said, "I'm only..."

"Who did you kill?"

"What? No one," Regi said, and the soldier holding him shoved him to the ground.

"Then where did you get these boots?"

"My brother-in-law's brother gave them to me," the words tumbled out of Regi's mouth, faster than anything he had ever said before. "He has a friend, a soldier at the military base in Xela, who sent them for me, I'm on my way to ag school, I didn't have the money to buy boots."

"Liar!" The soldier kicked him in the side. "Where did you get these boots?"

Regi repeated his story again and again, as the blows fell like rain, as the soldiers threatened his life with the kind of insults that only came before an execution or a fight to the death. Before long he began to wish that he was lying, that he could tell the soldiers something, anything to make them stop and let him go. He could feel the blood pooling around him, sticking his clothes to his skin. But he couldn't change his story, he had nothing to tell but the truth.

"I have the telegram from the school. It is a government school, I am not a rebel," Regi kept repeating.

"Stop," the soldier said at last. "Get up," he said to Regi.

Regi didn't know where he found the strength to comply, his whole body hurt, but he was able to stand. He waited, swaying, as the soldier looked him over. He felt as though any minute he would collapse.

"Alright," the soldier said at last, "maybe you're telling the truth. We're going to turn around and count to ten. When we turn back, we better not see you."

Regi nodded, he had bitten his tongue and it felt too thick to speak. None of the other passengers looked at him. The soldier started to count. Regi began running. He was slow, he couldn't trust his legs, but he ran. The fields were clear, the closest cornfield was too far away, the ground was plowed, there was brush on the side of the road—that would offer some cover, but not enough. The soldiers wouldn't bother to shoulder their guns, they would shoot from the hip. Regi knew his best bet was just to put as much distance between him and the men as possible. The ground had been harvested and the ground hoed so it was level, but his feet sunk into the loose soil with each step. Regi felt as if he was running in slow motion. He could hear the soldier's counting like the beating of some fatal drum. He remembered Alfredo's story, his brother laughing at the soldier's bad aim. The memory gave him hope.

When the soldier reached nine Regi looked for somewhere to hide; on ten he threw himself to the ground. Shots rang out over his head as he crawled across the open gap and made it into the cornfield. The stalks wouldn't stop bullets, so he focused on getting out of sight.

He crawled in and then to the left along the rows. Regi kept crawling long after he heard the gunshots stop, after the bus rolled away. He hid under a bush and just breathed.

The sun was high in the sky when he came to his senses. Blood crusted his skin and cracked when he tried to move. He was thirstier than he had ever been in his life. Regi staggered to his feet and went in search of water. He looked at the world through swelling-shut eyes and wandered directionless through the void. Across the fields he came to a farmhouse and knocked feebly at the door.

Regi was surprised the farmer didn't turn him away for fear that he was a rebel. He told his story once on the doorstep, and again sitting on the edge of the farmer's bed as someone cleaned the blood from his face with a damp cloth. A water glass was pressed into his hand, and Regi drank deeply. Not long after that, he fell into a deep sleep.

When he awoke the house was empty, the farmer was tending to his field. There was a little cloth packet on the table next to the bed with a few tortillas, and enough money for a bus ride the rest of the way to Villa Nueva.

He checked his pocket—the telegram was still there so he made a start at once, caught the first bus to the city, and then the one to Barcenas. He made it to ag school by the end of the day. Then the real work began.

Green Man listened as Kukulkan spoke, "There was a time when humans and beasts lived assss one with nature, they made sssacrifice to the Sssun God and didn't take more than their share. But now the balance is all wrong, humans sssteal from the Earth and give nothing back. The balance must be ressstored."

"But do you have to eat everything?" Green Man said. "You're a most impressive serpent, surely you could talk to the humans and help them live in harmony with nature."

"Humans cannot sssee me any more. They don't tell the old ssstories."

"I can see you," Green Man said.

"But you are different," Kukulkan said, "you are not human."

"I was born to humans," Green Man said, "that has to count for something."

"You don't live among them, you are more a creature of the forest than a creature of the village."

"Then why couldn't Vulture see you?" Green Man said.

"Vulture eatsss the humansss garbage. She feastsss on the excess of the ssspecies," Kukulkan said.

"Hmm," Green Man said, "I've taught people how to live in harmony with the forest, and many of the beasts who live here are my friends. What if I took your message to the humans?"

"It isss too late," the serpent said. "The Sssun God, who causes all thingsss to grow, hasss decreed that they mussst die."

"But you are nowhere near the village," Green Man said. "Why are you eating so many innocent animals?"

"I got turned around," Kukulkan said, embarrassed, "I don't wish to ssspeak of it."

"I see," said Green Man, who didn't. "Very well, I propose a deal."

"What sssort of deal?" The serpent sounded suspicious.

"You say that your Sun God is the source of all life, and I'm certain that nothing could escape your great mouth," Green Man began.

"Yesss," Kukulkan said, "It isss known."

"Well then it's very simple. I—who came from humans—will allow you to eat me, and if I can make it out alive, then you will admit that humans can be educated, not just slaughtered."

There was a great rushing noise, and it took Green Man a moment to realize that the snake was laughing. "You are a ssstrange little creature," Kukulkan said, "but a brave one. I will accept your terms."

The coils that held Green Man loosened, and Green Man was able to stand and get a good look at the creature that had bound him. Kukulkan had destroyed his home; its timbers were splintered around his feet. The serpent had green scales, but behind his head sprouted a rainbow of vibrant feathers. His coils stretched through the forest as far as Green Man could see. He felt a moment of fear as the serpent reared his head and yawned wide, but quick as lightning the world was lost in dark and wetness. It was all Green Man could do to scoop a handful of earth before he was tossed into the air, and the serpent swallowed him down.

Green Man felt himself falling, the slick walls of the serpent's gullet tight around him. And still he wasn't afraid. He crushed the clump of earth in his hand and felt the sleeping spores of a thousand tiny mushrooms burst into the air. Green Man closed his eyes and let his senses unfold. He could smell the spores, feel them on his skin. They were pinpricks of light in the darkness and he urged them to grow. When they brushed against the serpent's throat, the spores took hold, and began to grow, to spread. They multiplied much faster than any natural fungus—doubling, doubling in size and numbers until the serpent's throat was coated in ballooning growths.

Green Man reached out and caught hold of the mushrooms. Some crumbled under his hands, but his fall slowed and eventually he was able to hold fast. By touch alone Green Man could identify a dozen different mushrooms in the ever-growing carpet: this one was good to eat, that one could be dried and turned into a powder that would keep wounds from festering, that one could pull the poison out of bee stings. He turned his attention to each one in turn, feeding it energy, channeling the life force of the rainforest.

It was dark in the serpent's throat, and wet, and though the serpent's body tried to fight them off, the mushrooms were quite happy there. And so they kept growing, sending out tendrils, firing off spores, and gradually they broke down the wall of the serpent's throat. Green Man swam through the mushroom cloud, and broke exhausted into the sun-dappled world of the rainforest.

"What have you done to me?" Kukulkan cried.

"Shh..." Green Man said, "lay still and I will help you, but you must promise to abide by your bet and leave this place in peace."

"I will, I will," Kukulkan said. He twisted onto his side, and exposed the wound in his throat.

With a grunt, Green Man lifted a fallen tree trunk from the forest floor and set it on the wound. The serpent hissed in pain, but sensed Green Man's good intentions and kept still.

Green Man closed his eyes and followed the spores home, willing them to latch onto the rotting log instead of the serpent. As quickly as they had emerged, the mushrooms dried and fell off, blossoming again on the log in Green Man's hands. He smiled as the last of the mushrooms left the serpent.

The wound was not large compared to Kukulkan's great size, and so Green Man cleaned and dressed it. "Thank you, Green Man," the serpent said when he was done.

"Of course," Green Man said, "it's no more than I would do for any injured creature. You will tell the Sun that humans are not beyond saving?"

"I give you my word," the serpent said. Then he reared up, nearly blotting out the sun, and the feathers behind his head flared out. With a great leap the serpent took to the sky, and was gone.

Green Man looked around at his ruined garden, his trampled home. "I guess I'll be eating mushrooms for dinner," he said to no one in particular. Then he shrugged, and began to rebuild.

Chapter 6
Green Man's Gift

Green Man was sad. The rains hadn't let up for days, turning the floor of his house to mud, carrying away his garden, and keeping his friends all well undercover. Green Man didn't stay inside because he feared the rain—the water rolled off his skin like a thick frond leaf, and he could swim against any current. Snakes didn't dare to bite him, and if one did by mistake, he knew how to cure its venom. No, Green Man didn't stay inside because he was afraid, he stayed in because there didn't seem to be any point in leaving.

Sometimes he wondered if he could find a tree tall enough to climb above the clouds. But the relentless grey streaking rain left him feeling dull and tied to the ground. Besides, Green Man thought, what could possibly be up there? Just the occasional soaring bird with places to go, and better things to do than talk to me.

And so he sat at his window and watched for the sun.

For dinner Green Man ate the potatoes that hung in his house. He rolled them in salt and shoved them into the hottest part of his little fire, and let them cook until the skins were crisp and the inside was soft as butter. He ate them without tasting.

Even with his belly full, Green Man felt hollow. He took his net from the wall and tried to mend it, but his fingers wouldn't move when he told them to. He took his machete and whetstone from their places and held them, one in each hand, watching the dying fire play in the steel. Then Green Man put them away and climbed into his hammock.

He tossed and turned and tried to sleep, but dreams resolutely failed him. All he could hear were sheets of rain on his roof, and the rolling rivers of mud beyond his walls. Green Man imagined that his house was a tiny island in the middle of endless water. He imagined floating away from everything he knew, and appearing on some foreign shore. Animals bounded around him and Green Man felt welcomed, until they started to point and laughed their shrill alien laughs.

Even still, Green Man preferred their company to the empty heartbeat of the rain.

When the calling started, at first Green Man couldn't believe his ears. It was a mournful wailing, something ghostly, somehow ancient. Green Man found tears starting at the corners of his eyes.

He stood up and crossed to the door. The crying cut through the roar of the rain, and Green Man felt a tugging in his chest. The sound was like nothing he had ever heard before—deeper than the calls of the howler monkeys, more full of sadness than the wailing of humans.

Green Man stepped out into the rain and let the lethargy wash from his limbs. The water was cold but Green Man took no notice. His world shrunk to the size of the song and that insistent call drew him through the forest.

Rumor had it that a few years earlier one of the students at the Agriculture School snapped under the pressure. He brought a machete to a lecture hall and started attacking students left and right. How many he killed was a matter of much whispered debate. Some students said he killed over twenty, others that he was tackled to the ground before he killed anyone. Of course none of the students now had been there at the time. The professors wouldn't confirm anything, but there was one who walked with a limp, and confided in his students one day in the field that his left heel had been cut off "with a machete." After a few weeks Regi was certain that it had happened.

Fieldwork began every day at 6:30 in the morning, and students spent the next five and a half hours tending crops, fish pools and fruit orchards of every kind, milking cows, tending pigs, and managing pastures. A lot of the work was done by hand, but they also learned to operate combines and tractors and how to irrigate and spray chemicals. Most of the training was carried out on the hundreds of acres of farmland that the agriculture school owned, but sometimes the students would venture down to the large farms on the south coast, fields operated by the Universidad de San Carlos' Agronomy School. Regi found himself behind the wheels of machines that he had only seen from a distance; he felt the power thrumming through his hands and it made his whole body shake as he worked more land than he had ever imagined.

At noon they broke for lunch. Regi recognized the food that the cafeteria served, but it never tasted right, somehow both bland and over-seasoned. There wasn't much conversation over the lunch tables, just the sound of chewing. Sometimes a student would fall asleep in his plate and one of his fellows would give him a shake and everyone would share a brief commiseratory laugh before turning back to their steel trays. After lunch the students packed into various lecture halls to take notes on the biochemistry of agriculture and the science of soil from one until five in the afternoon. After a break for dinner they were required to study until ten each night.

Regi struggled, as he always had, to keep himself focused on the specific assigned readings that he had to get through. There was so much in the school library, and so many concepts that the assignments just touched on, that he would often find himself buried in the stacks tracking down the answers to some obscure question of germination or plant immune systems. He could finally figure out what that fungi was on his Dad's avocado tree, or why the tangerine trees on their farm never produced more than a handful of fruit.

While his grades suffered for it, his tendency to read wider than his fellow students quickly earned him a place as an out-of-the-box thinker. So when the day came to register for extracurriculars, it surprised no one that he eschewed the longer line of boisterous, broad-shouldered types trying out to be bullfighters and instead threw his lot in with the marimba club.

His fellow marimba players were serious, bookish men. They tended to look down on the bullfighters as all body, no brains. The bullfighters, of course, tended to treat them as though they were soft.

They all tried their hands at bull riding during the cattle ranching and horsemanship unit. Turning down the chance to ride didn't reflect well on a student, so Regi put on the chaps and hat and went through the moves. He absorbed every bit of advice that came his way, from where to put his hands to how to lock his knees or roll with a landing. He saw the doors of the bullpen swing open and then he blacked out. He came to as the students were cheering and the ranch hands corralled the bull out of the ring. Even if he couldn't remember his ride, the other students could and it earned him a measure of respect. Regi decided to stick with the marimba.

The conductor was older, greying, half-bald, with steady hands and a mustache that danced as he played. He wore a pair of small glasses that he tucked into the breast pocket of his shirt when he read. He looked a lot like most of the rest of the marimba players at competitions or rodeos—an aging breed clinging to the music they grew up with. He also served as the forestry professor and went by the name Valle Dawson, which Regi thought was weird—who gave their child two last names? It was only later when he read some of the books the professor had written that Regi learned his first name was Horacio.

Regi and his friends enjoyed playing music surrounded by men three times their age or more, and they played with an energy and passion that compensated for their lack of experience. The weekends that he spent traveling with the marimba club packed in a hot bus with his mallets crossed atop his drawn in knees, the rumble of the road setting up a dissonant clamor from the tightly packed instruments—they did a lot to help him stay sane.

On weekends when there wasn't a rodeo or a festival to play at, Regi would hop a bus into Guatemala City. The ride was dusty and uncomfortable and he usually spent it daydreaming. His schoolwork began to infect his fantasies, and so Green Man ran through row crops and marshalled swarms of beetles to take care of the weeding. He would call up rain right on time, and finish the harvest faster than any combine.

The worlds he lived in blended at the edges. Here were neatly ordered rows of corn and wheat and coffee, but in his mind he saw all around great trees hung with vines, avocado trees bursting from the fields, wild chickens ran through the rows. And here came Regi, a giant in the garden, spraying pesticides from his hands, banishing the natural order of the world, asserting his dominance.

The bus hit a pothole and Regi slammed his head into the window. The visions melted away like fog before the sun.

Guatemala City rambled out into the country, an endless maze of cement block houses and high office buildings. Metal gates gave tantalizing glimpses of courtyards hidden from the street, and bars bolted across windows made it feel like hooded eyes were watching Regi everywhere he went. No one looked casual on the streets; women walked home from the markets as quickly as their legs could carry them. Men kept their eyes down as they went about their business. Even the boys boasting gang colors seemed nervous. There was a studied quality to the way they lounged on street corners or leaned against the cool stone of alley walls. Soldiers patrolled the city in numbers, young men from the mountains with itchy trigger fingers, and claustrophobia from the too-close street.

Still, when the bus stopped at the station Regi got off and walked. He wasn't terribly afraid of getting mugged, he had nothing to steal.

Most weeks he spent the last of his money on the ticket into town, and if he had some left over it went toward pens and paper, and every once in awhile a small treat from one of the stores that sprawled along the school's fence.

Regi's destination was always the same. One of his sisters, Arge, was a nun in a convent on the north side of the city. The convent was a place of peace and stillness, a welcome escape from the mountains of work that greeted him most days.

The sisters soon got to know him, and they let him through the gate with a blessing and a smile.

"How are you?" Arge would ask, "Hungry?"

"Always," Regi said with a smile, "How are you?"

"There are tamales in the kitchen," Regi's sister said, and hugged him. "Eat something, and tell me how school is going."

In between mouthfuls of tender tamale, Arge would share news from home. Regi's younger brother Rogelio had gone on to study a career and the next younger, Rene, was studying for the ag school as well. "You've paved a path for the younger ones to follow."

Regi would relate the news of his week as well. He was always able to do well in one of his classes at a time, and so he would change his focus whenever he slipped too far behind. If he wasn't keeping up in genetics, it was because he was working ahead in soils. Arge would listen to his stories and shake her head.

"I don't know how you juggle it all," she said.

"I have to," Regi said. "It wouldn't be so bad, but I'm taking odd jobs at night to pay for everything, well, most of what I need at least."

"Do you need money?"

"No, no," Regi was quick to say, "I'm fine, I just haven't been sleeping much."

"I'm amazed you're still on your feet," his sister said.

Regi swallowed, "Well, there's a German businessman who has been staying at the college. He watches us in the fields some days. Rumor has it that he may give some students scholarships."

"Do you think you could get one?" his sister said.

Regi laughed, "He'll probably give it to someone with good grades."

"I'll pray for you," Arge said, and then they talked of other things. Regi asked about life back home, and his sister filled him in with the few details she had—her only contact with the village was an occasional telegram, or word from a traveling friend. At least she would assure him that all was well, and that their mother sent her love.

They drank coffee and walked the cool halls of the convent, talking or just being in each other's company. The silence helped Regi breathe and his tension slipped away unnoticed. When the sun began to dip he sighed, "I should get back to the school."

"Do you have the money?" his sister asked.

"I'm fine for the bus to the school, but I'll have to walk to the station."

"You will need to walk through half of the city, it's getting late and walking is not an option at this time of day. Here…" she reached into a pocket and handed him a few bills.

"It's too much," Regi said, forcing a smile. This was a ritual with the two of them.

Arge counted out just enough for the city bus to the station and thrust it toward him, "Take it."

"Thank you," Regi said.

His sister walked him to the gate and they lingered there a while. Regi knew he had to beat the setting sun to the bus station, but it was always hard to leave the convent.

Back at the school he sank onto his narrow bed. The dormitory walls were rough, made of red bricks, and moonlight played across the floor. The windows were wide and tall, glass panes held at the ends by aluminum frames that could turn to let in the breeze. Regi slit open a letter that had been sitting in his mailbox—no return address marked it as an internal communication.

The letter was terse, only two lines.

"Regi, please see the counselor about your grades. Your appointment is for 12 noon on Monday."

He spent the next day and a half in a haze of anxiety. For the first time in his life he struggled to eat, and when he found himself in the counselor's office he fell into the thinly padded chair feeling half dead, prepared for the worst.

"Are you sick?" the counselor said.

"No," Regi said, his tongue felt thick.

"You look terrible."

"I'm sorry."

"Why are you… Never mind," the counselor looked at her hands, they were clean with neatly squared nails, "There's no shame in struggling. Of the students who we accept into each class, three-quarters will drop out. We're proud of that number. This is the most rigorous course of study in the country. Those who graduate are the best and the brightest, and it falls to them to keep the people of Guatemala fed.

"What is at stake here is literally life and death, feeding or famine. So you understand why we can't go easy on you."

"I do," Regi said.

"That being said," the counsellor continued, "we do want to see you succeed. If you're struggling for health reasons, we could give you some time to recover, something like that."

"I'm not sleeping," Regi said, "that's all."

"Do you know why?"

"I don't have time."

"If it's just classwork, there's not much I can do…"

"I'm working," Regi said, and he told the counselor about his bus ride out of the mountains. He told her about growing up with nothing and walking for kilometers through the rainforest on an empty stomach. He told the counselor how he would carry tortillas to camp mid-week in a bag woven from plastic strips, and how his father would pack that bag for his return journey with as much of the harvest as he could carry. Regi told her how he dreaded carrying anything round, like pineapples, because they would rub between his shoulder blades and give him terrible blisters by the time he made it home, "And so I have to work in the evenings," he finished, "just to buy books and things."

The counselor had leaned forward across her desk while Regi spoke, now she relaxed. "Well," she said, "your teachers all have good things to say about you; creative, engaged… I won't make you any promises, but I'll see what I can do for you."

Regi saw the German businessman three times before the end of the week, and on Friday he got another summons to the counselor's office.

The counselor was all smiles, "I've got some good news for you," she said, "good news, and 25 quetzales a month so long as you maintain passing grades."

For a moment Regi just stared at her, expecting it to be some joke. Then, "You're serious?"

"I am. I passed on your story and, well, it made an impression."

Regi felt fire draining out of his body, through his fingers, across the floor and gone. He relaxed against the back of the chair. *What a difference*, he thought, *25 quetzales can make.*

The counselor was smiling, "Now, I vouched for you. I'm counting on you not to make me look like a fool."

Green Man walked for what felt like hours. Rain whipped across his forearms as he pushed aside waterlogged vines, unseen fronds clubbed him about the head and shoulders. Even he could hardly make out what was around him, let alone where he was. Each tree was just a darker shadow against the night.

Still the crying drew him on, a lasso wrapped firmly around his heart. Green Man knew he was close when he felt the song well up around him, filling the rain-cut air and making every leaf and every living thing glow. He hurried, stumbling like a drunk over roots and slick mud, until he broke into a clearing.

There the rain pulled back and the sun broke forth, and one single beam found its way through the leaves to strike the shattered heart of a hormigo tree. The tree had been struck by lightning and split in two, and the two halves had fallen apart in a great V. The wood was two-toned—around the outside was pale, almost like pine, but its core was the deep red of blood.

Green Man ran his hand across the wood and the cry changed. Still full of sorrow, the song no longer seemed lonely. Green Man felt as though he had found something that understood him, and he found himself smiling even as tears mingled with the rain streaking his face. Ignoring the wind, Green Man climbed. Grabbing here and there, he began to teach himself the language of the hormigo tree.

Gradually the storm faded, and with it the ghost-song. Still, Green Man felt as though he could sense it, trapped in the wood of the tree. He walked to his home, a much shorter journey in the light of true morning, and retrieved his tools—a machete and a heavy hammer.

Green Man worked deftly, striking a cleaner line with his machete than most men could make with a chisel, and piece by piece he freed the voice of the tree.

The voice came in a series of short boards, of different thickness and weight. Each resonated with a small part of the song that haunted Green Man's memory, and he strung them in order as best he could on a frame made from blown-down branches. He worked beyond hunger, past exhaustion, wholly consumed by need to recapture the song, and when he was finally done, he could hardly stand.

Still, Green Man needed to hear the voice of the tree before he could let himself sleep and so he ran his knuckles across the boards. His heart leaped to hear the sweet music, the ghosts of ages past crying out in pain and joy. The sound carried him far away from his round house and his little patch of garden, and he wept as it faded. But Green Man went to bed at peace, knowing that he had experienced something wonderful.

That was many, many years ago, and since then Green Man has taught hundreds to build and play the marimba, and so the voices of the long lost Maya have returned to the world.

Chapter 7

Green Man and the Garden of the World

Green Man and Vulture were talking one evening, as the sun began to sink into the rainforest canopy. Green Man was tending to his garden, and his friend watched him from the roof of his round house.

"I'm talking about ambition, Green Man," Vulture said. "What do you want?"

"I don't know what you mean," Green Man said, "I have plenty of food, I'm surrounded by my animal friends, I have my health… What more could I possibly want?"

"So this is it then," Vulture said, "Every day from here on out will be exactly like this one. You'll tend your little garden, you'll cook your meals for one, you'll throw some chicken guts on a rock in the hopes that I'll fly by and share a little conversation."

"Well…" said Green Man, "I suppose I would like to have a family, but I've never met anyone like me. No matter how well my garden grows, no one from the village wants to be a part of my life."

"Have you asked them all?" Vulture said.

"I don't have to, I'm not from that world. They think I'm some forest creature, like I'm a monkey. No one wants to spend time with a monkey, so I'm afraid you'll have to keep coming around to keep me company." Green Man smiled to show he meant no harm.

"You run with the monkeys. Surely you can find a friend as crazy as you," Vulture said. "You're a healthy young man, don't give up on human society so easily."

"I may be healthy," Green Man said, "but I don't know about young. I grew up in a matter of weeks, who knows how long I have left."

"Does that really worry you?" Vulture said.

"No," Green Man admitted, "I feel good."

"I think you'll live as long as the rainforest," Vulture said.

"Well then, I'll tend my garden forever, and you'll never want for company," Green Man said.

"Maybe," Vulture said, and ruffled her feathers.

"What do you mean?" Green Man said, and he turned to take in the wild life all around him.

"You've traveled far, Green Man," Vulture said, "but I've traveled farther, and I've seen things that make me wonder about the health of our world."

"Like what?"

"I've seen great fields of trees all slashed to the ground so that cows can roam, only for the cattle to be slaughtered in massive numbers and loaded into trucks and carried halfway across the world. I've seen clear-cut acres planted with nothing but coffee, worked by farmers who are treated little better than slaves. The soil is worked again and again until it is drained of all life and they spray it with every kind of poison to keep bugs, weeds and fungi away. They are so afraid of their crops showing any sign of 'imperfection' that they kill everything else—to the point that nothing wants to live there anymore. Of course, then when people eat the fruit from these fields, they don't know that they are eating poisons too. They wonder why they are always sick, always weak and don't think that it might be this perfect-looking food that they eat.

"I've landed in those fields when the farmers leave them behind, and the soil is so hungry that it barely stays together. It washes away in the slightest rain, damming the rivers, choking the fish." Vulture shook her head, "The rainforest is large and strong, but every day it gets smaller."

"How can they do that?" Green Man said. "Don't they know that the rainforest is a home?"

"They don't care," Vulture said. "They think this is the only way to make money, to support their families, to make a better life. They want things, Green Man. Not everyone is as easily satisfied as you. Once they clear the forest they plant pastures and contract the cattle and the operation gets bigger, and they make more money so they need more land... I'm starting to think that only some catastrophic failure will put an end to it all."

"Well," Green Man said, "maybe I want something too."

"And what's that?"

"If they want to take my home then maybe I should return the favor. I'll turn the whole world into a giant garden, and make all the food free for people to eat. I'll teach people to grow their own food so that they don't have to buy from big farms, so that those farms stop destroying the rainforest." Green Man stood up straight, and his eyes flashed like distant lightning.

Vulture laughed, "I hope you do. So long as you still find time to throw me some chicken guts. I'll miss our talks if you get too busy."

It was glorious, being able to sleep. Other students complained about the workload, but Regi threw himself into his studies. Everything fascinated him, not just for the science itself, but because he knew that somehow, somewhere, he would find a way to do it better.

Even as he learned which chemicals to spray, and how long a given field could be farmed before it needed to be left fallow, his father's intuition reverberated within him. He and his fellow students weren't working within a natural framework; they were fighting against nature and there had to be consequences for it.

The months churned on and students dropped out, overwhelmed or simply exhausted. Regi took each one as a warning to study and work harder. Time began to bleed and blur together.

The marimba club grew tighter and tighter. Regi drummed his mallets against his desk to keep himself awake when he struggled to focus, and he learned to lean into the energy of his fellows. They spent nearly every waking moment together, in the fields or studying, or relishing the sheer emotional release of playing.

At the national marimba championships a crowd of thousands gathered to watch marimba bands play their finest renditions of Guatemala's folk songs, dance numbers and salon music. The heat in the open amphitheater made Regi's tongue stick to the roof of his mouth while he waited with the band. They watched grey-haired men coax mournful tunes from their instruments, and women with smoky voices sang rain over the mountains.

When it was their turn, Regi and his friends mounted the stage and tuned their instruments. Regi felt the weight of eyes on him, he shook out his arms and held his mallets ready, and the world slipped away.

Playing was a little like dancing; he knew the steps but the steps didn't matter. They played with energy and style and swept the amphitheater into one collectively held breath. They played an old world into being, unfolding like petals all around them, they played a raw throat opening, howling. They set down their mallets to a deafening silence.

Then there was a roaring, thunderous applause and whistles from all sides. Regi became lighter than air, and looked down on the world and saw that it was rich with possibilities, where colors were bright and plants grew freely. A world where no one starved or disappeared.

And then he blinked.

His bandmates were hugging him, the judges were huddled behind their table. Their director was sitting misty-eyed and watching something very far away. When the judges returned their score, Regi was proud and beyond happy that the band had won, but the real world had reasserted itself with a sickening lurch.

There were soldiers at the entrances, saying nothing, always watching.

The vision haunted Regi on the bus ride back to the school. There were so many problems in the world, how could one person fix them? The systems he was a part of were so huge and he… he was tiny. *What do we need?* He found himself thinking, *Rich or poor, soldier or farmer. Security? Shelter? Of course, but first food.*

Without food there was no energy, without energy a person couldn't have security, couldn't make a shelter, couldn't start a family. But how could the people who had the least access to it take control of their food supply *and* make it secure enough that they can have a life outside of the fields? His parents and brothers couldn't take a stand against the soldiers because they were too busy finding the bare minimums they needed to stay alive. There had to be a better way.

The thought was pushed to the back of his mind by the rigors of final-year tests, and the normal patterns of planting, weeding, harvesting…. The rhythm of the work made it hard to think about the future, but late at night lying in his bed Regi would paint pictures on the insides of his eyelids, of thriving gardens and well fed families.

As graduation neared, representatives from agri-businesses began to visit. They bought dinners for the students and made a great show of their gleaming cars and pressed suits. The message was clear: "We've made it, and so can you."

 Regi got an offer from a chemical manufacturer to work as a traveling salesman. They offered him a company car, a salary larger than any he had ever heard of, and commissions on each sale. And Regi accepted almost without thinking about it, because these men didn't seem like the sort of people one said no to. He promised that he would go through a training period, and decide whether to sign a long-term contract or not when it was over.

At dinner one night Regi found himself wondering about the cars, the suits. Where did they enter the farming equation? Wasn't a farmer who bought chemical fertilizer from the company also buying some part of the fancy car? If that was the case then all he had to do was figure out how to replicate the results that the chemical company promised, using resources that farmers could produce for themselves.

He started thinking about cilantro, still the best he had ever tasted, grown in cattle manure and rocky soil. Of course not every farmer could also be a rancher, but there was something to the idea.

Regi pulled a notebook and a pen from his bag and started to do the math. How much land did a family need to sustain itself? How many cows would it take to keep it fertile year after year?

No, the answer was obvious, *not cows. Chickens.* Chickens were a more efficient use of land, cheaper to feed, and they ate bugs and weeds too! Regi stopped trying to work the numbers and started sketching fields. Rows of coffee planted in the shade of avocado trees. Corn and beans, tightly planted. The fields of annual crops would need a nitrogen fixer, but what?

"Hey," someone called, "Regi."

Regi looked up, the dining hall was empty except for him and a classmate hovering in the door.

"Are you coming?" his classmate said.

"Yes, of course."

Regi closed the notebook and followed his friend down the hall to the library.

"What were you doing?"

"Dreaming," Regi said. "What are we studying for tonight?"

Regi passed his final exams, and the week before the graduation ceremony passed in strange sort of haze. There were still weeds to pull, crops to water and harvest, but it all felt unreal compared to the fevered pace of the last three years. The chemicals company sent Regi a series of training packets that detailed their various products and he spent the evenings glancing over it.

It didn't sit well with him, the idea that all a farmer's problems could be sprayed away if only you paid the company enough, but the idea of a free car and a regular paycheck was too tempting. Every material thing that signaled success was available to him and all he had to do was get over the queasy feeling it gave him to take it.

It didn't help that his fellow students envied his offer.

"You mean they're paying you that much for *no* field work? You'll get paid to drive around the country in a car that they're giving you, using gas that they're paying for?! Some people have all the luck."

But Regi didn't feel lucky, he felt like he was taking advantage of something. Someone. By graduation time he still hadn't figured out exactly who.

The day came hot and heavy, fog flooded in across the fields and steamed off in scalding whispers as the morning wore on. The students sweated in their white robes while workers assembled the stage that they would walk across. Family members gathered in a grassy bowl below the stage, some mingled while they waited, some formed tight knots of stillness.

There were people from all walks of life and strata of society in the audience. Regi spotted a group of nuns, the Tau crosses prominently displayed around their necks. His sister wouldn't be the only one of the Sisters of St. Francis in attendance; most of them had watched Regi go through the daily struggles of the school and had come to celebrate. There were businessmen and farmers with black earth still ground into their hands. There were housewives, engineers and soldiers.

The German was in attendance to see how his money had been spent, and the head of one of the major agribusinesses gave the keynote address. The thrust of his speech, as far as Regi could tell, was that the graduates today would change the world by going to work for large companies and continuing to do exactly the same things that they had been doing for the past fifty years. He felt his eyes glaze over under the wave of self-aggrandizing platitudes.

It didn't take long for the students to receive their diplomas. Out of the one hundred and fifty students that had started school with Regi, only eighty had made it to the end and half of *them* had not passed the final test and were scheduled to keep taking it until they passed. Regi looked out into the audience as he crossed the stage, searching for his family.

He knew that he wasn't likely to find them. It was too expensive to travel. But there was Arge, beaming with pride and next to her… Regi almost tripped over his too-large robe. His father was standing with his sister, his clothes were still dirty from the road, but he was beaming and there were tears in his eyes. Regi waved, *Thank you Papa*, he thought, and swallowed the lump that started forming in his throat. He accepted his diploma, and left the stage with his head held high.

So Green Man took up his machete, his bag of seeds, and his hammock and walked off into the world. He ran across the dappled green of the forest floor until he reached a gravel road. Green Man paused just at the edge, and breathed in the feeling that he was about to make a choice from which there was no going back.

He rocked on the balls of his feet, closed his eyes, and turned his back on the forest. The cries of monkeys and the cawing of birds dissolved into a hum of truck engines. The murmur of human voices rose from the village like smoke on the wind, and the lowing of cattle felt harsh on Green Man's ears.

As Green Man entered the town he added a jump every few steps, so that he could see over the wood slab walls that surrounded the homes, and peer into the gardens behind. He was looking for a place to start his project, but he wasn't sure what it would look like.

He wasn't sure, that is, until he found it. The perfect place to start was a small garden planted in parched and rocky soil. What caught Green Man's eye was the care with which the meager yields were being tended; sickly beans were draped against round wood poles stuck at angles in the ground, rows of potatoes were kept meticulously clean of weeds and beetles. A watering can sat at the edge of the garden, a smile of rust painted across one side.

Green Man went straight to the door of the house. He knocked on the rough-cut wood door. "Who is it?" came a voice from inside.

"Green Man," Green Man said.

"Who?"

"I want to help you with your garden."

"What?" There was a sound of shuffling feet, and the door swung open to reveal a dried fig of a woman with bright eyes and a vibrant shawl, "What did you say your name was?" she said.

"Green Man."

"Your parents weren't very imaginative, were they?" The woman shook her head, "And you want to help me with my garden do you?"

"Yes ma'am," Green Man said. "It looks like you could use some help."

"Young man, I know how to garden."

"You put the work in," Green Man said, "I can help you harvest enough food to feed yourself."

"Whatever you're selling I don't want it." The woman began to close the door.

"Selling?"

"I don't have any money," she said, "Good luck, and it's a nice job on the costume."

"Costume?" Green Man said, "Wait, I'm not selling anything. I'll help you for free."

"What do you want?" The woman sounded suspicious.

"I want to help," Green Man said, "and somewhere to hang my hammock."

The woman pursed her lips. She looked him up and down. "Alright," she said, "Come inside."

Green Man spent the next few months helping to rejuvenate the soil. He sat in the garden until late at night and listened to the stories that the earth wanted to tell—of too much work and too much sun, of a deep hunger for living things. The soil mushrooms were almost gone. Green Man could sense a few spores here and there, but there were no underground systems to support them.

Green Man went into town and traded for a few hens and wood to build the shell of a home for them. He went to the rainforest and found wild-growing coffee bushes, round sticks for the walls of the henhouse and palms to cover the roof. Each day he spent long hours with the old woman, tending to the earth, or helping her weave the bright colors that she sold at the market.

The rainy season came and went, and the old woman's garden was bursting with food. Their harvest was so generous that they shared some with the neighbors—all of whom were eager for Green Man's advice on their own gardens.

And so the year ran out, and the people of the village ate better than they ever had before. The children put on weight and ran freely through the streets. The adults found it easier to smile and laugh. Until one day none of the men were to be found.

"Where has everyone gone?" Green Man asked the old woman.

"Up to the grazing fields, I shouldn't wonder," she said, "It's about that time of year."

"What are they doing?" Green Man asked.

"Clearing new pasture," the woman said. "Old ones have been all churned over, can't reseed them until the cattle are moved somewhere else."

"Thank you," Green Man said, "I think I'll go and see if they need any help."

"You run along then, I'll be here."

Green Man raced to the edge of the forest and, sure enough, there were men with roaring chainsaws bringing down the great old trees. Monstrous yellow machines hoisted and stacked the timbers.

"What are you doing?" Green Man shouted over the roar of the machines.

"Our jobs," one of the men replied.

"But you have plenty of food to eat," Green Man said.

"And we've never been healthier. Clearing the new pasture is going faster this year than ever before, and it's all thanks to you." The man's smile slipped as he took in Green Man's grim expression, "Is something wrong?"

"Everything I showed you, how to grow food, how to stay healthy, it all depends on working with nature, not against it. Can't you see that this runs completely against everything that I taught you? You're destroying the thing that would sustain you, can't you see that?"

"Maybe, Green Man," the man said, "but we get paid to do this, and so we can buy clothes for our families, and school books for our children. There's plenty of forest left."

"Here maybe, but every village cuts down more and more of it, grows cattle that are shipped off to who knows where and never give back to the land. This taking and taking can't go on forever." Green Man looked out at the men, they had stopped working to hear him speak. He could see softness in their eyes; they had heard him.

"We know," the man said, "but what choice do we have?"

"Alright everybody," a strange voice called out, "we aren't breaking for another two hours, what's going on here?" A light-skinned man pushed his way to the front of the group and stared Green Man in the face, "What the hell are you?" he said.

Green Man couldn't speak. He walked past the stranger, through the group of villagers, over the corpses of trees, and vanished into the forest.

When he was many kilometers away Green Man sunk to the ground and wept.

"I won't say I told you so," Vulture said from a nearby branch. Her voice was kind.

"They didn't listen, they didn't learn anything."

"You mustn't be so hard on them, or on yourself. There are systems in place that are bigger than anyone can overcome alone, even you. The only way to fight it is with systems of your own. We can't just change a few people's minds; we have to change the way the world works."

"I should have smashed their machines," Green Man said, "scattered the cattle, upturned everything."

"All that would have done is taken money out of the hands of the people you want to help. It's not about the individual machines, it's about the factories, it's about the people at the top who own the land, who control the flow of money."

"Money," Green Man spat, "What is that? Just paper and empty promises."

"It's a tool," Vulture said, "and like a machete it can give life or take it away. It all depends on how it's used."

"So I wasted my time."

"No," Vulture flew down next to him, "food is the key to freeing people from the systems they are trapped in. You can't force them into a new model overnight any more than you can grow corn in an afternoon. Whether you can see it or not, you've planted the seeds of your world-garden. Cultivate it, and it will grow.

"Remember, Green Man, that food is the one thing we need everyday. When the food we eat sickens us instead of nurturing, when the natural systems that are responsible for producing food are destroyed, and when more people are starving and dying even though their bellies are full, we will need a different system. If nobody works to create one, we will all be doomed. So don't take no for an answer, keep your focus on the horizon, and you will succeed. You have to."

"It doesn't look like that to me," Green Man said.

"Ah, but you don't fly do you?" Vulture said, "So I may be able to see farther."

"I hope so," Green Man said.

"Hope doesn't grow food," Vulture said. "There's a lot of work still to do."

Chapter 8

Green Man and the Soldiers

Green Man heard them a long time before he saw them. They were clumsy in the rainforest, not used to the twisting roots, the tricky dancing light. He could smell the fear-sweat on their necks, the gun oil on their hands.

Soldiers, he thought with a sigh, *again.*

It seemed that there were more of them every day, scaring the animals of the forest with their swearing and strange smells. Sometimes Green Man wasn't sure if there had always been so many of them, maybe he had just come to notice them more as he grew older—how different they were from the resting state of things, how intrusive.

Usually when they came to his part of the forest, Green Man laid low and waited for them to pass by. If they were headed toward his clearing, sometimes he would slip through the trees and make horrible noises—demon calls and tearing sounds that made even heavily armed soldiers turn away from their path.

But something was different about these men. Green Man couldn't say what at first, but there was something that made him raise his head and stare off toward the source of the noise. He couldn't see the men, of course, there were too many trees in the way, but he could track them by sound, by smell. There were at least a dozen of the men, a little over a kilometer away. They weren't running, but they had been not long ago… Are they lost? Green Man wondered.

If they were, then it was his duty to lead them back to the road—he owed it to the rest of the creatures of the forest to get them out of here as quickly as possible. His garden could wait. So he sheathed his machete and started off through the forest.

As he ran Green Man let his senses unfold and tried to get a read on the forest. The animals had scattered, that was no surprise, and the ferns cried out all trampled under boot and cut low by machete's swing. But there were other people moving through the trees as well, a wide arc on the far side of the knot of running soldiers.

Of course there were rebels in the forest. Green Man didn't worry much about politics, but generally the rebels knew how to move through the forest without disturbing every creature for miles, and they didn't set gasoline fires in the brush. The sound of gunshots was agony on Green Man's sensitive ears and he furled his senses in so that he didn't feel the rainforest's pain.

There was more gunfire, cries of pain and fear. Green Man dropped to the ground, disoriented. He dug his fingers into the soil and felt the vibrations of running feet. A hundred meters away a body fell and didn't rise, Maybe they'll sort this out for me… Green Man thought, but then the bile rose in his throat and he reproached himself for the thought. All life was sacred, didn't he know that? All people and beasts were a part of the web of life around the world, and harm to one was a harm to all.

Green Man pushed his will through the soil, and ferns burst up, growing rampant and wild, surrounding soldiers and rebels alike. He could feel the men's confusion as they found their paths blocked, shots spoiled. Then, taking a deep breath, Green Man called out in a voice like summer thunder, "Stop shooting, drop your weapons. If anyone else dies today I will be most displeased."

"Are you God?" he heard one quavering voice say. Soldiers and rebels alike complied with Green Man's command.

T he car didn't kick, it didn't splutter or backfire, it didn't cough turning over. The car hummed. Jets of cool air blew from the dashboard as Regi cruised through the height of dry season heat without breaking a sweat.

He was still technically in training; he told himself that he hadn't signed any commitments yet. That this just gave him the flexibility to find work that mattered. Then he leaned back into a seduction of leather and padding and lost himself for a while.

The work quickly became a routine. Regi would roll up to a farmhouse somewhere in Guatemala, say an hour north of the capital city where the land began to get rough and overgrown Mayan ruins were hidden by swelling hills. The farmhouse would be set back from the road, far enough that the dust cloud from Regi's car wouldn't quite reach the porch.

The houses were mostly long, single-story affairs, white to stay cool in the sun, with corrugated metal roofs. Regi would be noticed from the road and met at the door by the farmer's wife or father.

"Hello there," Regi would begin, and then smile, "how are the children?"

The trick was to get in the door, and to get the conversation started after that it was simply a matter of letting the conversation come around naturally to food or the weather. Regi found that farmers tended to trust him implicitly—he was still farm-boy skinny, with rope-like muscles and a deep tan. They confided their growing troubles to him.

"It's too wet. I can't get rid of the rot on my squash."

"My soil is no good this year; my corn is two weeks behind the season. I've given it plenty of water, more than I can afford."

"I moved south last year. I can't seem to get my tomatoes to grow here."

And for every complaint, Regi would go to his briefcase and pull out a thick catalog.

"What I would use..." he began, and flipped to the page with the prescribed chemical or designer seeds that the farmer could use. Nine times out of ten he'd closed the sale by the time the sun had begun to set, and usually he got invited to supper on top of it.

But those dinners rarely sat well with Regi. He didn't think that he was cheating the farmers that he sold to—he knew how the products worked and had used them at ag school. The part that bothered him was the part that sealed the deal—his personal endorsement. The more Regi studied the more he agreed with his father's intuition, that farming that was good for the earth was good for people and that farming that damaged the earth was harmful. The products Regi sold were all short-term fixes—they increased yields this season at the expense of future ones. He was careful to warn his customers that they would need to leave their fields fallow between applications of certain chemicals, but more often than not the people he was selling to didn't have the luxury of long-term planning.

Regi sensed that what he was selling and how he was selling it didn't fit. How could he tell people that these chemicals would help them grow more food when they were harmful, sometimes fatal if inhaled or swallowed? He couldn't shake the vision of himself, a half-dressed child, eating fruits straight from the field. How many of these people were eating poison disguised as food?

Just around the time the town's agronomist had introduced Regi's father to herbicides, his oldest brother, Everardo had left to work for the government's National System to Eradicate Malaria. The family had been desperate for some income, so despite all the concerns Everardo left to join his cousin and the other young men who sprayed homes and water sources with DDT. Everardo had worked evenings at a mechanical shop in town and had learned to drive and carry out basic repairs. These skills earned him the job of DDT transport and distribution, which meant loading and unloading the truck daily, bare hands and all exposed. Everardo landed at the hospital a few months later and was incapacitated for weeks. Regi had an image burned in his mind of his older brother laying listless in the hammock under the walnut tree in front of the house, eyes fixed and fighting to breath normally again. Everardo being the first born male was an anchor of the family, uncles and cousins came to visit, sometimes it looked like a funeral with coffee and sweet bread included, brought by the visitors. At least there were treats and food to eat, but Everardo just laid there in the hammock or in the bed apparently unaware of what was going on around him. Even as a young boy Regi understood what had happened as far as the chemicals having done this to Everardo. It frustrated him that nobody knew how to heal him. His mother's herbal medicines seemed to have no effect. And then, here he was, some years later promoting chemicals that held similar dangers for unsuspecting farmers.

When the chemical company asked him to sign a contract, Regi shook his head, "I just don't think it's a good fit," he said.

"But your numbers are good, you're making sales."

"It's the travel… I'd rather be working with my hands."

They didn't fight him too hard, there were plenty of other young men eager to sell for them. Regi decided to head home. He needed to clear his mind and reconnect. When he arrived in Poptún his father wasn't home, but his mother hugged him tightly, "He'll be glad to see you," she said at last.

"It's good to be home," Regi said. They ate a simple meal that evening, rice and beans and hand-tossed tortillas. It had never tasted better.

Regi didn't stay home for long. There was a Catholic mission in the southeast near the border with Honduras that served a town every bit as poor as the one Regi grew up in—only there the suffering was compounded by near-constant drought. The mission needed an agronomist and asked one of Regi's friends from ag school to come teach. The friend declined, but he passed the word on, and Regi answered the call.

Regi met with the head priest in Guatemala City after an introduction from the sister superior at the convent and was dispatched right away. He drove a truck with the priest, a Spaniard with squinty eyes, who was the interim director at the mission's school and smoked a seemingly endless stream of cheap cigarettes.

The cross-country drive was long and twisting. The roads, where they were paved, were full of potholes deep enough to swallow a whole tire. The rest was a dry, rocky dirt or gravel track. Every so often a car coming the other way would flash its headlights at him to let him know that there was a checkpoint up ahead. The warning gave Regi time to pull his papers, and himself, together.

Regi knew how to deal with checkpoints. As a child bringing supplies to and from his father's field he was stopped by soldiers who suspected him of running ammunition for the rebels. They made him unload his pack and scattered it across the ground and then watched as he struggled to pack it back together. He always carried more than he could easily stand up with, and so the soldiers would laugh to watch him like a turtle trying to turn over. Eventually one of them usually took pity on him and offered him a hand.

Since graduating from agriculture school Regi had an easier time getting past checkpoints. His agronomist ID and his ID from the chemical company were usually all he needed, and when the soldiers interrogated him he had perfected his line. He was an educator, he would tell them, doing his best to keep children in school and out of the hands of the rebels. If children stayed in school, they would be registered with the military and become productive members of society.

Of course, if it was a rebel checkpoint then it was a different story. Regi leaned on his roots as a child of farmers, one of the oppressed campesinos. He would tell them that he only carried government papers so that he could continue to help improve the lives of poor people across the country.

In ag school Regi had fallen in with a group of rebel sympathizers and learned all sorts of tricks for getting through checkpoints safely. He knew to say as little as possible because it was hard to tell when a checkpoint was manned by rebels and when it was manned by soldiers posing as rebels. It simplified matters that he had a priest with him—unless they had a specific reason to suspect a priest, people on either side tended not to bother them.

The stops were bad enough when he had a clearly documented reason for traveling, but he wondered what he would have done if he were on the road looking for work, or for some family emergency. The soldiers at the checkpoints mostly looked young and scared, and Regi knew that if ag school hadn't taken him the army would have come to his door eventually and then he would have found himself on the other side...

What would it be like, he wondered, *standing all night long, any car that passed might be full of people who wanted to kill you? Probably not much different from driving up to a checkpoint without a uniform on.*

Each time the car stopped Regi felt a gnawing in his stomach. While he waited for the soldiers to finish reading over his papers, he watched the cold steel muzzle of the gun pointing through his window.

When they gave him his papers back and waved them through he forced himself not to stare in the rearview mirror, waiting for them to change their minds. Waiting for the bullet.

When they made it to the mission in Jocotan he was exhausted and shaking. A nun from Belgium showed him to his room, and Regi collapsed onto the thin mattress without another word. The world went black with blessed relief.

The next day dawned hot and dry which set the tone for the rest of the year. The ground was cracked like alligator skin and Regi kicked up red powder in his work boots. The children in the classroom were all glazed over with hunger, their eyes were glassy and far away. It was a look that Regi recognized all too well.

The mission was staffed with volunteers and workers from all over the world, doctors and scientists. The Belgian priest had turned a school and convent into a massive compound with a hospital and an orphanage. Jocotan still had a mayor, but he held the position in name only—the priest ran the town.

It was no surprise, faith and hope were just about the only things the town had going for it. Perpetual drought left the population so listless and starving there wasn't even need for an active military presence.

At first Regi was excited to be a member of such a well-educated community and he threw himself into developing an organic agriculture curriculum. One of the Germans protested, "Are you kidding?" he said. "Look at the soil, look at the fields, these people need every advantage they can get. They don't have time to worry about chemical-free food."

Regi just shook his head, "It's because the environment is so hostile that we have to use these techniques. For years they've been trying to fight against the environment and look at what it has gotten them. Over time we can't beat nature. We have to figure out how to live with nature instead."

The German wasn't convinced, but he let the subject drop.

Still, the months of struggling to get anything to grow in the poor soil took their toll on Regi, and he was frustrated being surrounded by highly educated Europeans who tended to treat him like their water boy. Eventually he approached the priest.

"I want to go back to school," he said. "I can't do that here. My sister is helping to open an orphanage in Chimaltenango; I'd like to go help her."

"They couldn't pay you," the priest said.

"All I need is room and board. It will give me the chance to study and apply to the university."

"We appreciate having you here," the priest said, "but if you think this is what you need…"

"I do," Regi said.

Green Man demanded that the soldiers send him their leader, and he opened a path through the ferns so that the representative could find his way. The soldier only came up to Green Man's chest, and when he saw Green Man, he instinctively reached for where his gun normally hang—but of course he was unarmed.

"Why are you in my forest?" Green Man said.

"Y- you're green!" the soldier stammered.

Green Man stared at him.

"We heard there was a rebel camp back in the hills," the soldier said at last. "We were ordered to find it."

"I see," Green Man thought a moment. "Are you in charge?"

"We were ambushed, the sergeant was killed," the soldier said. "I drew the short straw."

"If I can get you and the others out alive. Will you promise to tell your superiors that the camp was destroyed and never come back here again?"

"Of course," the soldier said, "anything. I'm never coming here again anyway." He looked at Green Man again and shivered.

"Alright, go back to your friends; make them promise too."

The soldier swallowed hard, turned, and vanished into the maze of ferns. With a wave of his hand, Green Man closed the passage behind him. The rebels were in the stronger position, it would be harder to convince them to let this end peacefully.

Green Man jumped and snagged a tree branch. He swung himself up and up until he could look down on the whole mess of soldiers and rebels. From above they didn't look so different—just a lot of lost men wandering through the forest. Silently Green Man picked out his first mark, grabbed a vine from the tree, and swung down behind the rebel.

On the ground Green Man took his time observing the man. He wore a vest and tattered pants and a floppy-brimmed hat. He had a machete at his side, the handle was broken and held together with tape. He carried a gun that looked like it had been taken from one of the soldiers.

Green Man reached out and tapped him on the shoulder. The rebel spun around and Green Man caught his gun in a fern and held it tight. The man reached for his machete but stopped when he saw who it was.

"I heard stories," the rebel said, "but I didn't think they were true."

"I hope they were good ones," Green Man said. "I've spoken to the soldiers. They promised to never come back and to tell their officer that the camp has been destroyed."

"Don't believe them," the rebel said.

"Between you and me, I think they're too scared to come back, promise or no."

"It would be safer to kill them," the rebel said.

"Think," Green Man said. "They suspect you have a camp out here. If a whole patrol goes missing, they'll burn the forest to find it. If the patrol comes back and says they found nothing, that the sergeant was killed by a snake or a jaguar… That will take the heat off you."

"But if they don't keep their promise…"

"Make a deal with me and I'll help protect you. Keep trucks away from my part of the forest, keep your fires low, and don't cut trees you don't have to. Live in harmony with the forest and I promise no patrol will find your camp."

"You can do that?" the rebel said.

Green Man spread his hands to indicate the maze of ferns, "This was me on a bad day."

"Fine," the rebel said, extending his hand, "you've got a deal."

"The problem is," Green Man said, "that I don't trust any of you. Soldiers, rebels, whatever. So here's what we're going to do; I'll take all your guns and bring them to my house. Then the lot of you will work together to make a meal. Anything that you can harvest or hunt is yours to cook. Once you've eaten, I'll show you all how to get to your respective homes."

"What are you talking about?"

"And remember, I'll be watching," Green Man said, "so no funny business."

And in the end this is how it happened; the soldiers didn't know what they were looking for, but the rebels were native to the area and had grown up with the forest, so they were able to show the other men which plants were good to eat, and how to use a machete to dig up wild potatoes without *destroying the root. The soldiers were better fed than the rebels, and bigger, so they were better able to chop and carry firewood—they contributed lighter fluid to get the fire going.*

The men took turns roasting wild chickens on makeshift spits and burning their fingers pulling potatoes from the hot coals. They talked as they ate, and the more they talked, the more they found they had in common.

When at last the fire died down, and the food was finished, Green Man came to lead them home only to find that the men were reluctant to part. They wanted one last chance to say good-bye. They promised to meet again, under better circumstances. And then they consented to go their separate ways.

When Green Man made it back home it was very late, but Vulture was there, picking at the leftovers.

"I don't think I'll ever understand politics," Green Man said.

Vulture didn't say a word. She was busy picking through Green Man's trash heap for raw meat.

"We're simple creatures, you and I," Green Man said, and then lapsed into silence.

Vulture ate a piece of chicken intestine. Eventually they went to sleep.

Chapter 9
Green Man and the Strange Fire

It danced at the edges of his vision, flirting with the shadows of the forest. Green Man couldn't understand it; he'd never heard fire laughing before. And then there was the trouble of the color — now red, now green, and stranger colors like sunsets filtered through smoke or fog.

Normally Green Man was quick to investigate anything new in his part of the forest, but the fire was something completely different from anything he had known before. If it were burning out of control then Green Man could dig a trench to contain it or draw back the plants to starve it, but the fire didn't consume anything. It didn't grow, sending out tendrils that became sheets of flame, instead it moved like a human, but more graceful, like a jaguar but fearless.

Green Man found that he didn't want to look at it, didn't want to think about it. It unsettled the order of his world. And so he tended to his garden. He collected new fronds for his roof and tied them into bundles. He sorted through his beans to make sure that none were molding.

But no matter what he set himself to working on, the fire was there. It wasn't intrusive, but it was never far away. It wasn't threatening, but its strangeness terrified Green Man all the same.

Green Man built a fire to cook his evening meal, but he struggled to get it to burn as brightly and as cleanly as he wanted it. He kept feeding it more and more wood, but always that strange flame was brighter, and it made all other lights seem dull. Green Man continued to stoke the fire until it singed his eyebrows.

Recoiling, he shook himself and wetted a length of cloth that he wrapped around his forehead. Green Man had heard about fevers, and the strange things that they did to a person's brain, and since he had never been sick before, he could only assume that it was something like this. Green Man hung a pot over the fire and filled it with beans, onions and potatoes.

As his dinner cooked, he found that he couldn't smell it. He didn't stir the pot or tend to the fire. He was drawn again and again to the window, and each time there was the flame, quivering in the darkness.

Green Man ate without tasting and lay himself in his hammock. *I will sleep*, he told himself, *and in the morning it will be gone*. But he couldn't sleep, the forest was empty of its usual noises. No night birds hunted, or jaguar prowled. Nothing burrowed or scurried across the ground.

The fire has scared them all away, Green Man thought, *In the morning I must see what I can do about it*.

The morning took a long time to come, and Green Man considered many plans to deal with the fire. He would talk to it and ask it to please leave, or maybe he would fill buckets with water and throw them at the flame. He would go to the monkeys and they would weave a giant net to drop on the fire so they could drag it away...

Gradually Green Man fell into an uneasy slumber, and in his dreams he could feel the heat of the fire, taste its kiss like hot peppers.

When Regi arrived in Santa Apolonia, the orphanage was still under construction. A cluster of houses served as the core, where the children slept and ate. Across from them was a low building that would be the school, but in the meantime the orphans took their lessons on the tile courtyard from a harried looking nun named Sister Petrona.

The nuns lived in a dormitory within the walls of the orphanage along with the female volunteers. Regi stayed in the parish house in town and ate at one of the children's home. His room was furnished with a narrow bed and a cross on the wall, and Regi added little more—a pair of muddy boots, a stack of textbooks, and the guitar that he was teaching himself to play.

Arge greeted Regi at the gate with a warm hug, "Thank God you made it safe," she said.

"It's good to see you." Regi smiled. "It's quite the place you've got here."

"It is," Arge said, releasing him. "Let me show you around."

She took him through the gardens and brought him to meet several groups of children. They were well fed and mostly they seemed happy, but occasionally Regi caught a glimpse of something haunted in their eyes and it made him wonder how many had seen their parents dragged off, how many had seen them shot in front of them, or worse. These kids were casualties of war as much as any of the disappeared.

"What will I be doing?" Regi asked at last.

"Well that's up to you," Arge said. "In part, I told the priest you were an ag school graduate and he figured that you would know better than him what you could give to these children."

"Your gardens look good," Regi said, "I don't know that there's that much more I can do with them. So what else do you need?"

"So long as you can keep ten or fifteen kids occupied for a few hours a day, and preferably learning a skill or trade, you'll be a huge help." His sister tugged her sleeves down, it was the first nervous tick Regi had ever seen from her.

"Do you have a carpentry shop?" Regi said.

"We have an empty house and a bunch of donated tools," Arge said. "Let me show you."

Over the next few weeks Regi showed groups of children how to build a workbench and set up the shop; he showed them how to join bits of lumber and how to use the few hand tools he could gather from the community. One of the Canadian volunteers brought him a box of used power tools donated who-knew-when and Regi tore the old workbench apart to make room for a table saw, a wood turning contraption, and drill press.

With his workshop set up, Regi felt held back by a lack of lumber. Fortunately, the town's carpentry shop was just a few doors down from the parish house, and Regi could see them dumping scrap behind the shop in the evenings. The men who worked there were more than happy to let him haul the trash away for them, and if it went to the orphanage then so much the better.

The shop threw away the ends of boards and quite a bit of wire, so Regi drew up plans for hanging baskets that could be planted with flowers or lightweight vegetables and the children set to making them with a passion. The simple act of making something they could hold seemed to help draw some of the quietest children out of their shells, and soon there were planters spilling out flowers in every spare corner of the orphanage. Regi began sending the older children to sell the planters on the roadside, and that turned into a steady stream of income for the orphanage.

And so weeks turned to months. Volunteers came and went, from the city and from out of the country. More orphans came to them from burned-out villages and bloody hills.

Regi grew to distrust the foreigners that he worked with. Too many of them seemed caught up in their own troubles; they were doing the work to save their own souls first, and the kids a distant second. Volunteers from the United States especially bothered him; he had seen firsthand American guns in the hands of Guatemalan soldiers, and it was no secret that the U.S. was giving aid and support to the Guatemalan military with full knowledge of the massacres they were carrying out.

Regi wondered whether the Americans at the orphanage wouldn't be doing more good fighting their government at home, maybe then there wouldn't be so many orphans in the first place.

The other big group of volunteers was from Mexico, and too often these were people who seemed to view the children as a means to an end. They wanted to make contact with the revolutionaries and so they started a community quilting group and leveraged their connection with war orphans to gain the trust of revolutionary groups. Often they packed up and vanished overnight, leaving the permanent staff at the orphanage to scramble to cover their responsibilities.

Like any young man, a part of Regi wanted to go with them and fight to free his country from the dictator, but another part of him looked at guns like weed killer. *Is it possible to kill your way to peace?* he wondered. And so he spent his evenings in his room, walking his fingers up and down the neck of his guitar, or sitting with the sisters and a few volunteers and singing revolutionary songs. Regi never learned to read music, but he could hear the relationship notes had with each other and he followed them through winding melodies, into chord progressions and strange scales that were all his own.

One day Arge came to the house while he was helping one of his students shape the wire bowl of a planter.

"Can we step outside for a moment?" his sister asked.

"Of course," Regi forced a smile at the child he had been helping. "Nobody use the saws until I get back, okay?"

There was a chorus of agreement and Regi followed his sister onto the front porch.

"You're only using the east room, yes?" she said.

"I've been storing scrap in the west room—it can be moved" Regi said. "Why?"

"We've got a new volunteer coming in. She'll be teaching dance classes and the west room of this house would be the… most convenient place to put her." Regi's sister looked troubled.

"What's wrong?" he asked.

"It's just… she's another gringa, she'll take advantage of you if you let her and then leave you behind."

"What are you talking about?" Regi had to laugh a little, his sister looked so serious.

"Just promise me, okay?" She said.

"Sure, yes. I promise I won't let some white woman come and take advantage of me." The idea still seemed preposterous to Regi, *Why would she want to?* he wondered. He tried to imagine what this gringa looked like that had his sister so concerned—some steel-haired Belgian? A frumpy New York housewife with an itch for the exotic? Chances were she barely spoke Spanish.

"Have you met anyone nice?" Arge asked.

What's eating you? Regi wondered. "I've met plenty of nice people. I couldn't hope to marry all of them."

"You know that's not what I meant," his sister said. "I worry about you."

"You don't have to," Regi said. *Sometimes I worry about me too,* he thought.

It wasn't that he hadn't had crushes, or met women that intrigued him, but Regi had a hard time thinking he could settle down with any of them. The trouble was that everyone he met seemed totally caught up in trying to dominate their environment—fighting for more things, more money. Sometimes it was cloaked in the guise of humility or generosity, but too often it seemed to stem from a deep desire for control. Regi saw the same instincts in himself, but he always tried to take a step back and look for ways to live with or in the world rather than seeking to assert himself over it. He knew that would become impossible if he bound his life to someone who didn't share the same goal.

Regi spent the rest of the day in amused anticipation. *How funny,* he thought, *if this gringa were the one person that could keep me out of the priesthood.* He laughed to think of the look on his sister's face and let the thought drift away. *Ridiculous.*

And the next day he saw her. She was tall and willowy with a great mane of red hair. Her skin was pale, but speckled with freckles like shadows on the rainforest floor. He only caught a glimpse of her that morning—through the open door between his carpentry shop and her dance studio. He tried not to think about her as he prepared for the children to arrive, reminding himself of the conversation with his sister.

When they broke for lunch, the woman sought Regi out and sat next to him as they watched their charges play in the courtyard.

"Hi. My name's Amy," she said offering him a hand. "What's yours?"

Her Spanish was accented, but proficient and Regi felt a charge pass through him when he shook her hand. *I can see why my sister was worried.* He felt himself staring and blinked hard, "Regi," he said. "What brings you to the orphanage?"

"I want to teach, to work with kids, but I didn't want to go straight through for a master's degree. I had studied Spanish, so this seemed like a good fit. Besides," she laughed, "I wanted to get as far away from Minnesota as I could."

"Sure," Regi found himself nodding, even though he wasn't sure where exactly Minnesota was or why someone would want to get away from it.

"What brought you here?"

"Me?" Regi said.

Amy cocked her head and smiled at him. Regi felt his tongue go numb and it took a great effort to speak.

"My sister helped found the orphanage, and it's near the university so..."

"Are you a student?" Amy asked.

"Not right now, I hope to be. I graduated from the National School of Agriculture last year," Regi said.

"So you're taking a break too?"

"I'm running the carpentry shop, just next door to you," Regi said.

"I saw," Amy said, and there was that smile again.

"So you don't want to be a nun?" Regi asked, hope caused his heart to beat faster even as he tried to stop himself from being so ridiculous.

Amy laughed, "No," she said, "definitely not."

"Well," Regi said, "I should get back to work."

"Okay," Amy said, she seemed confused by his sudden change in tone.

A part of him wanted nothing more than to stay and talk to her for the rest of the day. For the rest of forever. He sort-of grinned at her, but when he got back to the shop, he made sure the door between their rooms was firmly closed.

And so it went. When they crossed paths they talked, so Regi tried to make sure that he stayed away as much as possible. His sister's warning, and his own distrust of Americans found fuel in the questions that she asked about his life, the stories she told about herself. Regi realized he had no way to know what her world looked like, or if any of what she said was true.

Each morning Regi would make sure that the door between their rooms was closed so that he wouldn't see her, wouldn't be tempted to talk to her, but somehow during the morning the door always seemed to swing open. He told himself that it was the latch, that he should look into fixing it, but he always found some reason to put that off another day. Some reason that had nothing at all to do with the way that Amy moved or the sound of her laughing.

Besides, once or twice he was sure he had caught a glimpse of one of her older students opening the door. He knew the children giggled whenever he and Amy talked, and he wouldn't put it past them to try to manufacture something.

In the end it wasn't the children that brought them together, it was mischievous Sister Mimi with her crooked-tooth smile.

Everyone knew that Sister Mimi was due to renew her vows that year, and there had been some talk about who would travel with her to her home convent. Regi was nonetheless surprised when she knocked on his door toward the end of one morning's shop session.

"Sister," Regi said, "come in, come in. Do you need one of the children?"

Mimi shook her head, "I actually wanted to talk to you."

"Oh?" Regi immediately felt guilty, though he wasn't sure why.

"You play the guitar, right?"

"Yes?" Regi said.

"Would you play at my ceremony?" Mimi said.

"I don't think you really… I'm self-taught, I don't read music," Regi said with an apologetic smile.

"If I gave you a recording do you think you could learn it by ear?" Mimi asked.

"Probably," Regi admitted, "But I really don't play for people. It's mostly just a meditation for me, a way to relax, I'm really not..."

"Nonsense," Mimi brushed away his protests like cobwebs, "bring your guitar tomorrow, let me hear you play. I won't take no for an answer."

Regi promised that he would.

The next day he played for her on the porch two of his own meandering compositions. Mimi clasped her hands in joy and insisted that he play when she renewed her vows, and so Regi agreed. He didn't notice that music from Amy's class had stopped, or that he'd been playing for an audience of more than one.

Mimi's home was nestled in the foothills separating a line of mountains from the ocean. They arrived in the early afternoon, Mimi herself and a couple of the sisters from the orphanage, Regi, and Amy—who he hadn't realized was going until he saw her getting on the bus with everyone else. His sister had waved goodbye to the lot of them, sorry that she couldn't go with them.

The air was sticky in the hills, despite a steady breeze off the ocean, and Sister Mimi suggested that they go swimming before dinner.

Most of the group stayed near the shore, letting saltwater wash over their bare feet, enjoying the radiant warmth of the dying sun in the black sand of the beach. The waves were gentle in the lagoon and the water was clear.

Only Amy dove into the water and swam out, fearless. Her long arms devoured the distance and soon she had beaten out past the breakers. Regi watched her with wonder, swimming had always been a bad idea in the rainforest, too many venomous snakes lurked in the muddy pools and rivers around his village.

He felt a friendly elbow tickling his ribs, "Go talk to her," Sister Mimi said.

"What would I say?"

"Help," Mimi said, "I can't swim."

"Stop it," Regi said, and she laughed at him.

"Come on, your sister isn't here to wag her finger at you, go talk to the woman."

"She has a boyfriend," Regi said.

"She has an ex that she came all the way to Guatemala to get away from," Mimi said. "The coast is clear."

"She looks like she's having a good time," Regi said. "I don't want to stop her swimming."

Mimi rolled her eyes at him.

"Did she put you up to this?" Regi said.

"She didn't have to," Mimi said. "Now go."

Regi waded out as far as he could and waved to Amy; he could feel Mimi's eyes on him from the shore. Amy waved back and swam to meet him.

While he waited Regi watched the sky and tried to think of something clever to say. He tried not to think of how little they were both wearing.

"How's the water?" he asked lamely.

"Warm," Amy laughed, "and light. Swimming in saltwater is so easy."

"Do you live near the ocean?"

Amy laughed, "I did once, but now I live about as far from it as you can get, I've been swimming as long as I can remember. There's not much else to do during the worst of the summer."

"You don't worry about snakes?"

"Why would I? Back home I don't think we have any dangerous ones," Amy said. "I take it you haven't swam much?"

"No," Regi said.

"I could teach you," Amy said.

"I would like that," Regi said. "It's too bad the weather won't hold."

"What do you mean?" Amy looked up. "It's as clear as a day could be."

"Look over the mountain there," Regi pointed. "You see how the clouds are gathering, like a great herd of cattle. It's a sure sign there's a storm coming. I expect it will break by midnight."

"Where did you learn to tell the weather?"

"When you're in the rainforest, there's nothing more interesting than when the next storm will come. You learn the signs," Regi shrugged. "Of course, in the forest you can't usually see the sky, so you have to sort of feel the air. The change in pressure can tell you a lot... Of course we would watch the animals too, they're much better at paying attention than we are."

"Oh?"

"Yeah, when the monkeys shut up and take cover you know to hang your hammock extra high."

"Would you take me into the rainforest sometime?" Amy asked. "I would love to see it."

"Yes," Regi said, without thinking about it, "of course. When would you like to go?" And then he stopped talking as his brain caught up with his tongue and they started tripping over each other.

"I'll ask for a day off as soon as we get back to the orphanage," Amy said.

The light was failing and the sisters were waving to them from shore so the two went in. As they walked up the long hill to the church they found themselves falling farther and farther behind as they talked about their childhoods, about what they wanted from life, about everything and nothing at all. Regi was surprised to find that she was easy to talk to, and the time passed too quickly.

"I think," Amy said, "that it's important to make plans for our lives, but to make them full in the knowledge that we will fail and that not only is that okay, it's wonderful. If everything went according to plan, if everything was totally under control then it wouldn't be life. The beauty in the world is in the unexpected, the out of control," she laughed, "Maybe I'm just making excuses."

"No," Regi said, "that makes sense. Think of all the little failures we have to look forward to. Can I tell you something?" Regi asked.

"Go ahead Amy said," a bit confused.

"I have this dream ever since before agriculture school," Regi started, "to set up a farm somewhere in the highlands with waterfalls, windmills, crops and forests. It would be a place to learn a different way of growing food. One that does not kill anything, but is based on letting nature give all the life it can. In this farm, young people would come to learn so a new generation is raised that stops what we are doing to the world today."

They stood for a moment as darkness settled around them, not speaking anymore, not needing to. When they could no longer see each other clearly, they went inside.

A few weeks later Amy twisted her ankle while Regi was helping her sneak back into the orphanage after the gate was locked for the night.

She had come over to the parish house for dinner and to learn how to play the guitar. Regi made spaghetti and chicken meatballs, and after dinner they sat in the living room. The window was open and a cool evening breeze made the curtains dance. Regi got out his guitar and tuned it by ear; Amy sat cross-legged on the floor next to him.

"Show me what you remember," Regi said when he had finished tuning, and passed her the instrument.

"This is G, right?" Amy said, Regi nodded and she strummed through a simple chord progression.

"Good, good," Regi said, "only this finger goes here…" he cupped her hand to show her, "and you'll get a clearer tone if you drop your wrist down, like this."

"That's right," Amy nodded, and copied his hand with her own. She played through the progression again and Regi leaned into her, resting his hand on the rough wooden floor—not quite touching her, but there where she could lean back into his arms. If she wanted to.

"Oh, I don't feel so good," Amy groaned.

Regi snatched his arm back, *Oh no…* "What's wrong?" he asked.

"I just, oh, have got to get to the bathroom." She staggered to her feet and dashed for the door.

Regi was still cursing himself for moving too quickly, when he heard her vomiting in the bathroom. Then he managed to be relieved and concerned at the same time; he paced the length of the room a couple of times and then went to get her a bottle of water.

Amy was draped over the toilet looking miserable. "Thanks," she managed, taking the water bottle.

"What's wrong?" Regi asked.

She raised an eyebrow at him as if to say, *What do you think?*

"Food poisoning?" he said. "I'm so sorry, I thought the chicken was fully… Can I get you anything else?"

"A new stomach," she said, and shoved her hair back as she heaved again into the toilet.

Fortunately the nausea subsided, and she was able to keep some water down. Regi wrapped a blanket around her shoulders and helped her to the couch.

"I'll be fine in a few minutes," Amy kept saying. "I just need to lie down for a bit."

Once she was comfortable Regi sat on the floor, leaning against the couch. Amy hovered just on the edge of consciousness. "Have I ever told you about Green Man?" Regi said.

"No," Amy murmured.

"Would you like me to?"

"Sure," Amy said.

"Green Man lived all alone in the middle of the rainforest," Regi began, "in a round house made from the trunks of trees. His skin and his eyes were the green of wet leaves, but his hair was black as night…"

As he spun out the story, Amy stroked his hair. Regi closed his eyes and felt moonlight on his face.

It was nearly midnight when Amy felt strong enough to go back to the orphanage, so Regi walked with her. It wasn't far, but he wouldn't want anyone to walk alone after dark. The gate was locked of course, so they went around to the west side of the compound. On the other side of the wall they knew there was an open building where the nuns did their washing. It had a shallowly slanted tile roof and so Amy could lower herself from the wall to the roof, and from the roof to the ground with ease.

Regi boosted her up and she grabbed ahold of the lip of the concrete wall, braced her feet and hauled herself over. When she was safely on the roof on the other side, she poked her head back over, she had a bright grin on her face.

"Thank you!" she whispered.

Regi waved, and she disappeared. He waited, ears tuned to the sound of her feet on the tile. When he heard her slip and stifle a cry, his breath caught in his chest. He heard her land, hard, and began backing up take a running start at the wall when she called out from the other side, "I'm okay, I'm okay."

"Goodnight..." he said. "Feel better."

"I'll do my best," Amy replied.

The next morning she was limping. That was the day the sisters started teasing them about their "friendship."

Green Man woke with a start. The flame was in his house—its light filled the single room with dancing colors. He reached out and touched it and found that it didn't burn but washed over his hand pleasantly.

"What are you?" Green Man asked, climbing out of his hammock.

I am a part of you, *the fire seemed to say,* just as you are a part of me.

"I don't understand," Green Man said.

You don't have to, *the fire assured him,* no one does.

"Why are you here?"

To show you that you are not alone.

"So you won't leave me?"

That depends on you, Green Man.

"I don't understand."

Open yourself to me, and you will never be alone again. Stay fearful and I will vanish without a trace.

"How?" Green Man said.

That's something you have to discover for yourself.

Green Man took a deep breath, and turned inside himself. He hadn't thought that he was afraid before, but at his core he found something cold and tight, curled in on itself. "Is this what you mean?" he asked.

Yes, the fire said.

"Help me," Green Man said, and the fire reached out to him, reached through him, and touched that cold ball above his heart. Green Man watched himself blossom open like a flower in the sun. He felt warmth flowing through his veins.

When he opened his eyes, the fire had vanished, but he knew that it was with him, part of him. He felt more present in the world than ever before, he could speak the language of the wind and water. He could feel the heartbeat of the earth.

It takes work, *the fire sang to him,* to keep me burning, to keep yourself open, to love. It takes hard work.

"It's worth it," Green Man said.

Chapter 10
Green Man Among the Ants

When Green Man woke up he was somewhere strange. There was earth packed all around him, and he couldn't see. He could feel movement on all sides, thousands of creatures thundering.

Curious, Green Man patted himself down. He wasn't hurt, but he didn't have matches in his pocket or his machete at his belt. Something rushed passed him. *So there is a tunnel up ahead.* Green Man put a hand on the ground and tried to feel the shape of the creature but none of the vibrations made sense.

It was huge, at least twice as big as him, with six legs. Its skin was thickly armored and it felt the earth ahead of it with two waving probes. As it ran, it chattered its mandibles, jaws Green Man was sure were at least as long as his arms. Green Man had never seen anything like it roaming the forest; he wondered if he had been sleepwalking.

All of a sudden there was a vibration behind him and Green Man became aware of another one of the creatures bearing down on him. He grabbed ahold of the mandibles as they cut toward him and vaulted onto the thing's back. It was hard to find a handhold on the tough skin, so Green Man propelled himself forward as quickly as he could landing on his feet on the other side.

"What are you?" he asked, but the creature didn't answer, instead it started to turn toward him.

Fortunately, the tunnel was too small for the thing to turn quickly, and Green Man seized the opportunity to sprint the other way. There had to be some light somewhere in this warren, if he could only get his bearings, he could figure out what to do. Think.

Reaching out, Green Man tried to get a reading on the earth. Mostly it was tilled soil, rich and black, but here and there he could feel larger rocks, chunks of granite and... there, flint.

The creature was on his heels, six legs churned the earth behind him, feelers swatted his back. Green Man poured on another burst of speed down the tunnel, around a corner, and began digging into the wall. The thing caught up with him; he could feel it rear up and at the last second he leapt aside.

The creature's mandibles struck the piece of flint and a spark lit the tunnel like a flash of lightning.

"Ants?" Green Man said. Because that's what the thing was, its carapace was black and shiny, segmented into three parts, "How did I get here?"

And then the spark died, and the tunnel filled with a rushing roar of darkness. Green Man could feel the ant bearing down on him but this time he didn't fight it. He reached out and his hand met one of its feelers, and Green Man felt his experience bloom. He was one of thousands of tiny minds, dancing soul-sparks, all tangled up in one guiding consciousness. He understood that this ant didn't want to hurt him.

For all the insect's grasping mandibles, it was a worker not a soldier. It was under standing orders to keep the tunnel clear of debris. *I'm afraid we've had a misunderstanding...* Green Man pushed the thought into the hivemind and felt it ripple through the network. Something rippled back and the worker ant bent its front legs in a sort of bow. "Well now," Green Man said, "now we're getting somewhere."

He mounted the creature, locking his legs into the narrow place behind its head. "Where are we going?" Green Man asked, but the ant couldn't answer. It stood, and sped down the tunnel.

Playing guitar at Sister Mimi's celebration had opened new doors for Regi. Beyond the music lessons with Amy, Regi had began to share his talents publicly. Now that he was back at the university he had time to perform. He had started playing guitar with the band when he saw a flier posted on the university campus in his first week of classes. He remembered how playing the marimba had helped him through ag school, and he went to audition. Now they were being called to perform at local schools around Guatemala City. One of these gigs, landed them at a local high school music club, Santo Domingo Xenacoj on the road to the highlands, where they were welcomed as the featured performers. In between sets, the club manager came to the side of the stage. He motioned the band over.

"Not to worry you guys, but we just got word a military patrol is coming through town. They will likely raid us so things might get hairy. You know where the back door is?"

Regi nodded. It was the first thing they had showed them when they got to the club—standard operating procedure for a band playing revolutionary folk songs.

When Amy was nervous about his getting caught up in a raid and disappearing, Regi said he wasn't worried. "We're playing music. Why would they care about that? It's not like we're getting into *stuff*. We won't have weapons or anything. Besides, we're students," Regi shrugged. It was hard to communicate out in the villages, but near the university campus in the middle of Guatemala City the military felt a lot less dangerous. Since its founding, the University of San Carlos had been considered sovereign by the state and soldiers weren't allowed on campus, so naturally it was a hotbed of revolutionary thought.

"You know what happened the last time students disappeared?" Regi said.

"Their families got to hold funerals without bodies," Amy said.

"Word went out to students at all the universities and high schools in the country. In one night they all turned out onto the streets and marched on local governors, military bases. There was rioting for days, and even though everyone knew there would be no arrests, it had sent the message that the military couldn't act with impunity. If there's one thing in the country that the military is afraid to mess with, it's the students."

"I don't want revenge," Amy said, "I just want you to be safe."

"I will be," Regi said. "I promise."

"I've got a lookout on the street," the manager's voice brought Regi back to the present. "She gives the signal, I'll hit the light in the sound booth and we'll go to prerecorded music. That's your cue to grab your gear and go. We're burning the flyers already. All they'll find is a club full of people dancing to Mexican pop music."

"Thanks," the lead singer said, "we appreciate the heads up."

The band was two songs into their final set when Regi saw the light go on in the sound booth. He stopped playing and stood up.

"Thank you everyone," the singer said, mid-verse, "good luck."

They had to leave the drum kit behind, but they slipped backstage with the rest of their instruments. The singer made sure they were all clear before following, he'd been stopped by soldiers before.

They went down a concrete hallway with flickering overhead lights and out the back door into a courtyard. They were careful not to run, no sense attracting unnecessary attention. On the other side of the building they heard a door being kicked in.

"Down on the ground, on the ground!" the soldiers shouted.

One by one the band slipped down the alley. Regi held his breath, expecting to hear gunshots at any moment, but none came. Their van was parked two blocks away, there were no soldiers in sight. Once they were all loaded in and the door was closed, they began to laugh.

"Don't they know buildings have more than one door?" the drummer said. "That's what I've always wondered."

"There is more than one reason why soldiers aren't allowed on the university campus, that's all I'm saying," one of the horn players said.

"Can we get going?" Regi's said. The others acted like they were already home free, but Regi knew what it was like to grow up in the shadow of a major military base, where every family in town had someone disappeared.

"Yeah, yeah," the singer started the van and eased it onto the street. He rolled down his window and lit a cigarette. The moon hung low above the distant rainforests, casting shadows like grasping fingers.

The road out of town wound back into the hills and soon the village vanished from sight. The band members didn't talk much, now that the adrenaline of the escape was fading, and the only sound was the growl of the engine, gravel spitting across the road.

Then they rounded a bend and were face to face with a roadblock.

"Shit," the singer said, and pumped the brake.

"Drive casual," Regi said. "Don't do anything to make them suspect us."

"Right," he stubbed out his cigarette and rolled the van to a stop.

The roadblock had been pulled together quickly. A jeep was parked across the road, and three sleepy looking soldiers stood around it, rifles pointed at the ground. One of the soldiers stepped up to the van.

"Everybody out," he called.

The singer rolled his eyes and opened the door so fast it nearly clipped the soldier. Regi bit his lip and slid the side door back. He made sure his hands were clearly visible, he made a conscious effort to move slowly and steadily.

"Where are you coming from?" the soldier asked.

"We're students at the university," the singer said. "We can come and go as we please."

"I asked you a question," the soldier said, a sick smile spreading across his face. "Where are you coming from, where are you going to?"

The singer was about to speak, but Regi cut him off. "Please," he said, "I work with Aj Quen. We were looking for new artists and our van broke down. My ID is in my right pocket, I'm going to take out my wallet and show you." He was breathless, staring straight ahead. He had been brought up never to lie, and even this half-truth didn't come easily to him.

"Go ahead," the soldier said, licking his lips. Regi had seen his type before, the power had gone to this one's head. He was out for blood, Regi couldn't give him any excuse to find it.

Regi pulled out his wallet and handed it to the soldier. The man flipped it open and thumbed through Regi's IDs, read his travel rights, and came to the card that identified him as an employee of Aj Quen. He looked from the card to Regi, then back to the card. For a long while he didn't say anything. Regi knew that their lives hung on a coin flip in the soldier's head.

As he held his breath waiting, Regi found himself remembering a trade show a few months back. Exporters of all kinds had set up booths in a stadium in Quetzaltenango, displaying everything from coffee beans to farm equipment and t-shirts. Regi was manning the booth for Aj Quen and had set out a spread of brightly colored blankets, hand-carved charms and wooden bowls. By midmorning he had made good contact with several foreign buyers looking for folk art, and he was looking forward to lunch when the helicopter landed outside.

The native Guatemalans immediately tensed. Even though most of them represented businesses that were friendly to the government, no one was free from the threat of violence. When the president himself entered the building, flanked by his top generals, a silence settled on the fair so thick you could have cut it with a machete. Regi gripped the edge of his display table to keep from shaking while he waited for the president to work his way around.

When the party reached Regi's booth, he had almost gotten calm. It would be no use lying about what he was doing, there was a stack of pamphlets on the table that described the mission of Aj Quen in detail.

"What do we have here?" one of the generals asked. The president looked bored.

"We are Aj Quen," Regi said. "We work directly with artists in communities across Guatemala to help expose their work to a global market," he took a deep breath, his next sentence could be construed as treason. "We also help to organize small artist cooperatives to better leverage their limited resources and help them compete with industrial manufacturers." One of the generals raised his eyebrows, no one spoke so Regi kept going, "We've achieved a great deal of success already. If you look at our pamphlet, you'll see that we were the largest non-food exporter in the country last year. By revenue, at least." He waited; the president's face was unreadable.

"Take the pamphlets," the president said at last. One of the generals grabbed the stack off the table." Do you have any more?" the president asked.

"One box," Regi said.

"Take that, too."

And then they moved on. A few weeks later the organization was awarded a special presidential honor, but Regi didn't let that fool him. He knew that they were being watched closely for any sign of communist or rebel sympathies and that the government was using them to keep tabs on local organizers. At the first sign of trouble, they could be rounded up and killed, but in the meantime Aj Quen was being given more than enough rope to hang itself.

"You're still in violation of curfew," the soldier said at last. "We'll need to search the vehicle." He threw Regi's wallet on the ground.

"Go ahead," Regi said. "I'm going to pick up my wallet and put it back in my pocket." The soldier was no longer paying attention to him.

He motioned to one of the other soldiers, who held the band at gunpoint while the first soldier ransacked the vehicle. He stabbed through seat cushions, emptied the glove compartment and tossed the instruments around. At last he emerged with Regi's guitar, a saxophone, and a trumpet. "Are these folk art?" he asked.

"Those are our own instruments," Regi said.

"Good," the soldier said, and one by one he flung them down the hill. Regi watched his guitar vanish into the darkness, and winced to hear it bounce off of rocks with a snapping of strings and finally a hollow crack. "Now get out of here before I change my mind."

"Thank you," Regi said.

They drove awhile in silence, the singer fuming. "You shouldn't have done that," he said at last.

"If I hadn't, he might have killed us," Regi said.

"Then we would have died," the singer said. "Better that than to just bend over and take their…" he had to swerve to avoid a pothole. "You shouldn't have done that," he said again.

Regi didn't answer.

The next time he went to visit Amy at the orphanage, he had a hard time talking.

"What's wrong?" she said.

"Nothing," he said.

"Come on," she said, "I know you better than that."

Regi said nothing, Amy crossed her arms and gave him the look she gave her students while she waited for them to settle down. "I think I might be willing to live in the United States," Regi said at last.

"What happened?" Amy asked surprised.

Regi told her, but he left out his lie. He said that he wished she had been there, even soldiers and gangsters knew better than to harass an American or the people they traveled with. He didn't tell anyone how much it bothered him that in his own country she did more to keep him safe than he could ever do for her.

Regi ran out his scholarship at the university and accepted a full-time position with Aj Quen. Amy left the orphanage and the two of them rented an apartment in a town called Chimaltenango, and Regi helped her get a job at a partner crafts exporter called Pop Atziak.

They knew the arrangements were temporary. Amy wanted to go back to school, and Regi didn't intend to work in marketing and exports forever, but they didn't talk about moving again until Amy found a master's degree program that she wanted to pursue. The University of Minnesota sounded terribly foreign to Regi.

"I don't want to live in a city," he said.

"You don't have to," Amy said, "there's plenty of country in the Midwest, and anyway the program I'm applying to is only fifteen months long."

"Maybe I should stay here…" Regi said, "keep going to school. We can decide after we're both done where we want to live."

"I don't want to separate like that."

"We should get married first," Regi said, "so we know we'll be together."

"I couldn't do that without my family," Amy said. "Its different for me, we are still really close. My sister is my best friend, I couldn't get married without her next to me."

"A civil ceremony then," Regi said, "so that I can get through immigration cleanly and then we'll have a formal ceremony with your parents blessing."

"That could work," Amy said.

"You won't have to wait for me to wrap things up here," Regi said. "You'll be able to apply to schools as soon as we're married and I'll follow you as soon as I can."

"I don't want you to give up on your schooling," Amy said. "We both have goals, we can both reach them."

"I can go to school in the United States just as well as I can here. In the middle of the capital I may as well be living in a foreign country anyway."

Amy laughed, "I love you."

They held the civil ceremony with the town's mayor as the officiating authority and the director of Aj Quen as witness and celebrated with tostadas and fresh-squeezed orange juice. Amy left for Minnesota soon after, and Regi began to wrap up his business in Guatemala. His contacts with Aj Quen had secured him a multiple-entry business travel visa and he was ready to go when Amy gave the word.

"There is one more thing I would like to do with you before you leave," Regi said.

"What would that be?" Amy asked.

"You see, when I went to agriculture school, I had one single visitor the whole three years I was there, my older brother Everardo. He came without announcing himself on a weekend, he took me to a town nearby and bought me my very first pair of cowboy boots. All other memories from the school are precious, but his visit was my highlight over everything else that happened in that school."

"When I started talking about going to the agriculture school in middle school, teachers, friends and my own family just discounted it as a pipedream, something that others with "resources" could aspire to, not us. The trail was not cleared yet. Once I opened the trail to the agriculture school, my younger brother René followed. By then, my father was supportive and even delivered him to school. I grew jealous at first, and then realized what a blessing I could be if I used my entrepreneurial spirit to open new trails. I would like to take you to visit René."

They boarded the bus from Santa Apolonia to El Trebol where the roads part ways. One highway leading to the highlands and the other to the extensive flatlands of the south coast. There they took the local bus to Bárcenas, Villa Nueva, where the school is located. This time, Regi's heart pounded with anticipation and excitement, not from fear of the unknown, but because the memories that he had held until now, despite dozens of times he had traveled this road, were the memories from the admission testing days, from the arrival for the first time, beaten and demoralized. This time was different, he was going there to give something back. Regi felt that this would also heal parts of his heart where the resentment toward his father's abandonment still resided. But that was not all of it, Regi wanted to show Amy how he envisioned a school that thought differently, and what a better place than the school of agriculture itself.

René welcomed them at the dorms and showed them around, Regi took Amy back to building two, where he had spent his three years. Then they walked the extensive vegetable gardens, rows and rows of them, extending over two hundred acres. Next, the avocado orchards, the experimental groves, the fruit orchard with dozens of varieties of citrus and exotic fruits, the aquaculture systems, flower gardens where Regi had grafted thousands of rose stems. Then they went to the animal production units, the pastured beef, pork and chickens, and the confinement buildings where animals were cooped up."

"I have never seen animals penned up like this before; on my grandma's farm everyone ran free," Amy commented.

"I kind of guessed that," Regi said, "I brought you here because I wanted to show you what our farm would not look like, but also how a school is structured and how we could build one that teaches a different system."

When Amy was accepted to a fifteen-month intensive master's program at the University of Minnesota Regi already had his bags packed. Despite being robbed one last time on his way to Guatemala City he arrived in Minnesota on August 2, 1992 with two cases of woven blankets and assorted other crafts. Amy's former Saint Olaf professor and his wife owned a small plot of land just south of Northfield, and allowed Regi to live there in exchange for help in their garden while they figured out a more permanent plan.

The ant burst into a massive cavern. Green Man could feel movement all around him, the rattle of carapaced bodies all driven by single-minded intention. Here the web between the minds was tangible, so thick in the air that Green Man could taste it, like wet chalk.

The ant seemed to move more slowly through the cavern, but that was only because the open space was so huge. Green Man marveled at the architecture of the earth, pressed into place, cemented together without tools or machines. And then the hive spoke to him.

"Green Man," he could hear the voice in the vibrations of the earth, in the unified chattering of a thousand pairs of mandibles, "thank you for coming."

"Thank you for showing me," Green Man said. "This… this is incredible."

"We are dying, Green Man," the voice said. "We are barren."

"I'm sorry," Green Man still felt lost. "I didn't know. You seem healthy."

"We are the final generation. There are no more young. They are taken to soldiers, to workers. There are no more. We are barren."

"Everything dies eventually," Green Man said, "Perhaps it is…"

"Not you," the voice cut him off, "you keep burning. Share your fire, Green Man, bring us back to life."

"What do you mean?"

"You make plants grow, move, bear fruit. You can do the same for us."

"Seeds," Green Man said, "spores, yes. I can bring life to a fallen tree, but nothing can bring that tree back to life." There was a rustling all around him. "I'm sorry," Green Man said.

"You will not even try?" the voice said. "Then we are lost."

"Where is the queen?" Green Man said. "I'll see if I can help."

The bodies parted and Green Man walked between them. Even from a distance he could feel that the queen was dying. She loomed in the center of the cavern, a hulking carcass. All around her the ground was hard with dried birthing fluid.

"May I touch you?" Green Man said.

"Yes."

Green Man flattened his palm against the queen's abdomen. He closed his eyes, and felt his way through her. He found nothing that surprised him. The queen was not sick, she was old and empty. She had stopped giving birth because she had no more eggs. Green Man opened his mouth.

"We can see what you see, Green Man," the voice said. "You are right. Thank you for trying, but you should go now."

"I'm sorry," Green Man said again, taking his hand away from the queen, "but I don't know how to get home."

"Let your body die," the voice said. "It's only a grub."

"What?" Green Man said.

"The last child was a hard birth; we knew there would be no more. We have watched you for years, and at last we sent a scout while you slept. We guided your dreaming into the grub, and you shaped it to your consciousness the way we would shape it into a warrior or a worker…"

"Or a queen?" Green Man said.

"We do not know how," the voice replied. "We cannot see ourselves."

"I can see you," Green Man said.

"But there are no more children," the voice said. "It is too late for us, let us collapse in peace."

"No," Green Man said, "I have an idea, but you will have to let me help you."

"Tell us what to do."

"Let me know you. I will send that image back to you, and you map it onto the grub, the me," Green Man said. "You didn't make me, I did. If I can leave this body then maybe it will revert to its base state and you can make it a queen."

"It will replace us."

"Yes," Green Man said, "but the hive will live on. Life creates life, that is our most fundamental function."

The voice was silent; Green Man could feel the mind considering. "Yes," it said at last, "that will do. We will end, we will go on."

Green Man touched the queen again, and built up the most complete image of her that he could. Every impulse, every instinct. The need to spawn and to live. Green Man tried to hold the whole web of the hive in his mind and give it back to the queen. He felt the bottom fall out of his stomach.

"Thank you, Green Man," the voice said. "Thank you."

Green Man forced himself to let go of the body. Dying felt like falling, like sinking in muddy water. And then he shot to the surface, alone in his house. Carefully he got out of his hammock.

He never found the ant hive for sure, but he never forgot to look for it, or wonder at the beauty of a world he had never seen.

Chapter 11
Green Man Goes to Town

Green Man had a good idea of how much food he needed and planted his garden so that he always had enough—but not so much that it would go to waste. But nature often has other plans, even for Green Man, and one season his garden yielded so many fruits and vegetables that he didn't know what to do with them all.

Every morning he was out at dawn picking tomatoes before they fell and rotted, and he was amazed to find fruit maturing in a day. There was squash and sweet potatoes by the ton, and he could only shake his head in wonder at the height of his bean plants, how heavy they were with fruit.

At first Green Man thought that his good fortune wouldn't hold, so he harvested what was ready, ate what he could, and processed the rest for storage and seeds, but late into the season he was still finding new fruit and he was looking forward to the biggest bean harvest in years. He began to think about what he could do with the extra food.

Now the beans he could dry and store for a long time, but the rainforest sustained Green Man all year long and so he didn't think about ways to preserve his harvest, he just wondered what he could do with it all.

There was a rebel camp nearby where the men were often hungry, and one night Green Man snuck up unseen to leave armfuls of produce near their cooking tent. But the men thought it was some kind of trick and they buried the vegetables with the waste from their latrines.

Then Green Man tried leaving food by the side of the road where the villagers walked on their way into town. Some children found the pile of vegetables and took an avocado each which they peeled with their teeth and ate while they walked, but the rest spoiled and was only food for flies.

Green Man tried not to let that frustrate him. "Flies," he told himself, "deserve to eat. Ants deserve to eat."

But flies and ants always had plenty of food, it was humans that he saw going hungry.

Green Man was sitting in a tree by the side of the road when he heard two women approaching. It was late in the day and they were coming from the town.

"It seems like every week there's less at the market," one of the women said. "Soon it will just be tortilla sellers and trinkets for tourists."

"That means less competition for us," the other woman said.

"Maybe, but why does nobody sell vegetables anymore? Is the military requisitioning food? Are they getting better prices at another market?"

"Who knows?" the second woman said. "Who cares. We're doing alright."

"I suppose we are," the first woman said. "I just wonder..." and they passed away toward the little village.

"Well," Green Man said, "perhaps I could bring my vegetables to this market. Vulture is always saying I should figure out how to make money work for me, maybe this is what she means."

And so the next week Green Man built a cart and loaded it high with all the vegetables he could harvest—great stacks of green beans, ears of corn, piles of tomatoes and avocados, and he pushed that cart to market.

Traveling on the road made everything seem farther away, but Green Man made good time and he reached the town early in the morning—not long after the roosters started crowing. He parked his cart near some other people with displays of weavings, carvings, jewelry and, yes, vegetables of their own.

Soon a young woman with an empty basket walked up to the cart. She examined his wares and clucked her tongue. "How much for the beans?" *she asked.*

"How much would you pay?" Green Man said.

She looked at him, confused, "You're not from around here," she said.

"No," Green Man said.

"Thank you," the woman said, and drifted away.

What did I say? Green Man wondered to himself.

The experience repeated a couple more times, and Green Man turned to the merchant at the stall next to him. "Why does no one want my beans?" he asked.

"It's not your beans," the weaver said. "They think you're weird."

"Am I weird?" Green Man said.

"Yeah. Try giving them a price when they ask for it," but the weaver's eyes told a different story.

"I don't know what a fair price is," Green Man said.

"Don't worry, they'll haggle," the weaver said, "but you've got to play your part."

"Oh," Green Man said.

"When you came through customs, you said you were here to visit," the immigration official said. She said it slowly and carefully as though Regi might not have understood her the first three times. The interpreter didn't copy her cadence. Outside snow was falling, cars crept by along the slick Minnesota street.

"I did, because when I came I thought it was just for a visit," Regi also wondered if maybe he wasn't being understood. He was sweating in his second-hand sweater, "But I realized that I wanted to stay here after all. My wife is going to graduate school at the University of Minnesota…"

"You said that," the officer cut him off. Regi wondered why she got to repeat herself but he didn't. "Do you expect me to believe you got married in Guatemala, she moved here, and you came to this country thinking you might not stay?"

"Yes," Regi said, but sensed that wouldn't be enough. "I grew up in the rainforest," he said, "I lived in small villages all my life. I didn't like living in Guatemala City because it was too many people, too much removed from the land. I was afraid that it would be like that here." He wished he could show her how much his hands needed earth to work, food to grow. But how could anyone who grew up in this desert of concrete and glass possibly understand?

"And it isn't?"

"I would rather be close to Amy and far from nature than the other way around."

The immigration officer stared at him for what felt like an agonizingly long time, then she sighed. "Fine," and grabbed the passport. She wielded her seal like a knife, stamping a null over the multiple entry visa marker, and approving a temporary green card. "Come back as instructed when your temporary card expires. We'll see if you're lying then, and you," she said, turning to Amy, "Remember that you would also face a ten thousand dollar fine and jail time for lying to a federal officer."

"Thank you," Regi said, "Thank you." not quite understanding everything, but hoping things would be ok. Outside he turned his collar up against the snow that hit like little daggers while he waited for the bus that would take him back to the closet-like basement room that he and Amy shared in a neighborhood called Seward.

For a few months Regi had continued to work for Aj Quen. Minnesota was full of little towns each with their own summer festivals: Jesse James Days in Northfield, Gold Rush Days in Oronoco, Buffalo Bill Days in Lanesboro; he even traveled to Luverne, Pipestone, towns that a lot of Minnesotans themselves have never heard of. Regi made sure he had a booth at each one.

He would arrive early and get set up. He had brought cases of crafts with him to the U.S. and he took great care to arrange his tables in bursts of eye-catching color. Walking around the market street while the rest of the vendors set up, he knew that he stood out. There was pottery and woodworking on sale in other stands, and pastel paintings of sunrises and corn fields, but he had the market cornered on textile art.

Regi had always been good at making sales and he went into the first few festivals with energy and optimism. He called out to passersby, and smiled and nodded his way through conversations that he only half understood. He had prepared stories in English about the source of his wares but he struggled to connect with his audience, and he found his stall passed over again and again in favor of the amateurish crafts in the booths on either side of him.

Are you just not interested in anything different? Regi wondered on his long drives back to the basement apartment that he and Amy rented in Minneapolis. For the first time in his life he was in a position where it didn't matter how hard he tried or how well he thought he presented himself, he wasn't breaking even. It was an unfamiliar feeling for Regi and difficult to give up a dream. He had seen the craft market as a way to not only to continue supporting the artisans he had been working with back in Guatemala, but as a way to build capital to eventually purchase some land. All of the work so far had only built them more debt. By the end of the summer Regi realized that he needed to try something different, and so approached the Resource Center of the Americas in Minneapolis to find buyers for his remaining crafts and he started looking for other ways to make a living.

He eventually got an offer from the Seward Montessori School in Minneapolis to work as a teacher's assistant. Regi jumped at the opportunity to gain insight and experience in a North American school system. He had always dreamt of building an agriculture school, one that would teach an alternative to conventional agriculture, but that was a dream for Guatemala. In the United States he didn't have the slightest idea of where one would start; learning of how a school works, that seemed like something that made sense.

"I'll warn you in advance," the principal said before she introduced Regi to the student he would be assisting, "this kid, Matt, has gone through more than a few mentors. He's functional enough to be in a normal classroom at this point, but if he doesn't make some progress that's going to change. Right now none of the kids will have anything to do with him because he's a biter, he likes to scream, and he's way behind on verbal skills."

Regi shrugged, "Okay."

The principal shook her head, "You're a better man than I." She punched the button on an intercom, "You can bring him in now."

Regi watched through the frosted glass window pane as someone in a wheelchair was rolled toward the office. He opened the door and stepped aside. The boy was small, his head seemed too large for his body, and his fingers were long.

"Hello," Regi said. The boy moaned at him.

"He knows English," the principal said. "His comprehension is very good, but he's got a small vocabulary and doesn't like to use it. Hopefully you'll be able to help with that but we don't expect miracles."

Probably it doesn't help to talk about him like he isn't even there, Regi thought. "Alright," he said out loud.

"You read his treatment plan?" the principal asked.

Regi nodded. Amy had helped him translate the document, and he'd gone through it until he had it memorized.

"Do you want someone to help you through the first week or so?" the principal asked.

"No," Regi said, "Matt and I will be okay." He had read the reports left by previous assistants and marveled at the difference between their assumptions and his own. The problems seemed clear to him and he was fairly sure that having a third-party present would just spoil his opportunity to get to know Matt.

So they were shown to the classroom. It was like no classroom Regi recognized. Where he had expected rows and rows of desks, instead the students sprawled across the blaze orange carpet, playing with bead boards and strangely shaped blocks. *What the heck did I get myself into*, he thought.

"Regi?" the teacher said. "My name is Teresa, I'm so glad you're here."

"What's going on here?" Regi asked looking around the room in amazement.

"What?" Teresa said and then, taking in his bewildered expression. "Ah, you've never seen a Montessori classroom before… We believe that every child learns in a different way and at their own pace. We try to create an environment that fosters natural learning instead of using a traditional lecture-based model."

Regi nodded, *just like farming*, he thought.

"We keep students together for the whole day and teach multiple grades in the same room so that the kids maintain a sense of community and learn from each other. Of course, some students, like Matt, require more help than others. That's where you come in." Teresa looked at him with a mixture of gratitude and pity. "Here are his workbooks and flashcards," she said, and Regi nodded. "Good luck, and let us know if you need anything."

"Thank you," Regi said.

Regi sat down next to Matt and introduced himself. He got the flashcards out and explained what they would be doing. He could see Matt eyeing his arm and carefully ignored it. Regi retrieved the first flashcard from the box and showed it to Matt, "What does this say?" he asked.

Matt grabbed his arm and bit down hard. Regi tightened his lips but forced himself not to react. He wasn't bleeding, at least not badly. "What does this say?" he said again.

The boy bit down harder and Regi felt his skin break. He kept his face and voice under control and repeated himself again. At long last the boy let go of Regi's arm, "Angry," he said.

"Good," Regi said and set down the card. He took a cloth bandana from his pocket and wrapped his wound before drawing the next card or attention from the rest of the kids, "Now what does this say?"

The boy never bit him again. Over the course of the year the boy learned that screaming and hitting wouldn't get a reaction from Regi either, and so he stopped lashing out at other people, and Regi used the opportunity to expand his vocabulary while getting paid.

When the principal asked Regi what his secret was, he searched for the right way to phrase his thought process. "People," he said at last, "don't do things just because. They want something and they have learned that behavior gets it. The way to train that behavior out of someone is to teach them that it won't get them what they want—in this case it won't make me leave and stop making him do flashcards—and you also teach them a positive way to get what they want."

"Well it seems simple," the principal said, "when you put it that way." The school year moved forward, Regi worked with Matt by day and continued to look for avenues to sell crafts by night. Slowly, doors opened and Regi was able to transition the craft endeavor to work within a local organization, the Institute for Agriculture and Trade Policy. Meanwhile, Amy had finished the university and landed a teaching job in Saint Paul. So it was now Regi's turn to complete his studies.

Life was a balance of school and work. It was full and it was good. The initial immigration hassles of entering and settling in a new country were fading to a distant memory and all things felt achievable again. The agriculture school idea was back in Regi's thinking, a brief interlude in the city he thought, complete his college degree, find a piece of land, start a farm the way he had drawn it while in agriculture school, and eventually start a training school.

Another school year was ending when Amy came home with some surprising news. They were expecting a child. Regi would be a father! While the timing wasn't part of the plan, Regi was overjoyed. There was still one more hoop to jump before obtaining a permanent green card (permanent legal residency in the United States), but that was still seasons away and Regi no longer feared the system.

The week before Regi's immigration hearing Amy went into labor, and gave birth to their son William (Willito). Sleep-deprived and scattered, Regi missed his green card appointment and when he realized what it could mean he dialed the office with shaking hands.

"I'm so sorry, I'm so sorry," he said, "we have a new baby and I completely forgot…"

"You have a baby?" the receptionist on the other end of the line asked.

"Yes," Regi said. "I haven't slept in a week."

"Just come in tomorrow," she said. "We'll try and fit you in."

Regi arrived early in the morning, but there was already a line stretching from the formica desk to the door. Amy gave him a kiss, took the baby and started back to the car. Regi stepped into the line with a sigh.

Just a couple of minutes later one of the immigration officers came up to Regi and asked his name.

"Good," he said when Regi introduced himself. "Step over here please, we'll get you taken care of."

Apparently word got around about his story and the paperwork for his green card was set aside with a set of card-stock forms. Someone looking official came out to the other side of the counter. It took Regi just a few minutes to sign where he needed to sign, then he stepped aside again to get fingerprinted, then he was done, just like that. *I can learn to love this country*, Regi thought to himself.

Amy and Regi bought a house in Minneapolis' Seward neighborhood, not far from the school. It was a tall, narrow duplex painted a fading yellow. The first thing Regi did was sift his fingers through the earth in the backyard; to get a sense for the soil he closed his eyes. There were raspberry bushes growing thickly along the property line, they suddenly extended for acres in front of him; rows of oaks and a thick undergrowth of bushes separated the raspberry rows. He opened his eyes and he began turning the yard into a garden. The Seward neighborhood was also home to the Community of Saint Martin's, a place where Regi met the first people who understood where he came from. Jack Nelson Palmeier, Laurie Krause, David Gagne, Steven Clemens, among some of them. These folks knew not only the inside story of how the United States had been involved in creating the conditions Regi had lived under his whole life, but they also knew how it affected the communities and the families. Regi felt understood sufficiently to be comfortable, he knew this community—although not a farming bunch—was the kind of people who would support a transition from the city to the countryside and would rejoice in him finding a way to achieve his mission. They knew very little about farming, but they surely knew about community and building support systems to hold each other accountable and supported as everyone sought to build a world of peace, fairness and harmony. That was all Regi needed for now, the rest of the story still needed writing; and so he bent back on the ground and kept going.

Even with a plot of land forty feet wide and a hundred feet deep, Regi was farming again, planting beans and tomatoes, and it felt like coming back home. He had been too long away from the earth; he had missed being a part of the cycle of green and growing things.

The next woman to ask for Green Man's beans was tall and harried looking. "Three," Green Man said.

"Three centavos?" the woman said.

"Yes" Green Man said.

"I'll give you one."

"Thank you," Green Man said, and the woman filled her bag with beans.

A few more customers came and went and Green Man found himself smiling. Soon there was a line of people waiting at his stand and the man next to him was just shaking his head.

"What's wrong?" Green Man asked him in between customers.

"You're letting them rob you," the man said, "and the rest of us too."

"What do you mean?"

"Look around," the man said, "no one will buy from any of us so long as you're giving food away."

"I'm not giving it away," Green Man said.

"One centavo a kilo? That's giving it away. You must not have a family to feed," the man said.

"I grow enough for me. This season my garden has yielded way more than I could possibly eat. I tried giving it away for free but no one wanted it. They burned it or passed it by. This was the only way to not waste it," Green Man explained.

"That may be," the man said, "just be careful or you might find yourself run out of town."

"I'm almost out of food," Green Man said, "then I'll go."

On his way out of town Green Man heard running feet on the road behind him. He turned to see five farmers carrying machetes and looking angry.

"You owe us," the first one said. She was short with a thick neck and two missing fingers.

"Here," Green Man said and tossed her his bag of money. She caught it and stared at it.

"Not enough," the woman said.

"That's all the money I have," Green Man said.

"Hmm," the woman said, "you've got a green thumb, don't you?"

"Among other things," Green Man said.

"We'll start with those."

Green Man cocked his head, "I don't understand."

She raised her machete.

"Oh," Green Man said, "how about I make you dinner instead?"

"Too late!"

But Green Man was faster than them and he led them into the rainforest. One by one his pursuers gave up until it was just the one woman left. She chased him into his clearing and he stopped and turned. They had run a long time and she was too tired to raise her machete.

"Welcome," Green Man said, "to my home."

"What are you?" she panted.

"I'm Green Man," he said. "Let me make you dinner."

He made fresh tortillas and served them with slices of avocado and chicken cooked tender in tomatoes. The farmer ate in silence.

"I'm sorry," she said at last. "We are so hungry, and we needed the money."

"It's alright," Green Man said, "I've got plenty of food."

And so for the rest of the season he brought his extra produce to the farmer's house and planted the seeds with her family. He taught the little *boys to climb trees and forage for wild fruits, to coax chicken into traps, and to watch for patterns in the weather.*

Chapter 12
Green Man and the Long Drought

The rainforest turned brown that year. Green Man even noticed the tips of his fingers starting to wilt and a russet fog crept across his fingernails. The monkeys came to him and asked for water, so he helped them dig deep holes, but even the underground rivers were drying, leaving behind hard grey soil.

"I'm sorry," Green Man said, "I don't know what's happening."

Even his connection to the forest thinned and faded to almost nothing. The rattle of dry leaves was an unfamiliar language, and all the trees cried out with overwhelming thirst. Each day Green Man walked out in search of water and each day he ranged farther and farther afield.

One night he ended up alone on the mountains, his arms wrapped around himself shivering in the biting wind. He pressed himself back into a hollow not far from the road and tucked his knees into his chest, but still the wind seemed to come from all around him. Green Man rattled like leaves and he felt the wind trying to tug away pieces of him. He clamped his mouth shut and waited for morning.

At his feet, Green Man had two buckets of water that he had filled from a stream a few hundred meters, a good hour's climb, above him. The stream shot out briefly from a rock cropping and then vanished again below ground. Green Man had run his hand through the grasses that grew where the stream disappeared, trying to track its path, but the strange plants didn't know him and wouldn't share their secret roots with him.

Huddled in the dark Green Man found himself drifting toward sleep, but instinct warned him to keep his head up. Somewhere in the dark space between waking and dreaming he heard a voice on the wind, or in the wind.

"Who is there?" Green Man called, and the voice came again, but he couldn't make out the words.

"Who is there?" he asked again, and this time the voice responded "I."

"Who is 'I?'" Green Man said, standing. "Where are you?"

"I am all around you, Green Man," the voice said clearly, and it seemed to reverberate through the stones and the ground at his feet. "You drank of my blood and took shelter in my stones, but if you would see my face, it is above you."

"It is dark," Green Man said.

"That has never frightened you before," the mountain replied. "The climb will warm your limbs and talking will do you good. I see why you fear, perhaps I can help you."

"Can you bring back the rain?" Green Man said.

"Come up," the mountain replied, "and we will talk."

So Green Man picked up his buckets and began to retrace his steps up the steep mountainside. With his back to the wind Green Man did feel warmer, and it was good to be moving again.

Before long he passed the outcrop where the water still ran, and not far beyond that he found himself face to face with a vertical wall of stone. "How do I move forward?" Green Man asked.

"Climb," the mountain said.

So Green Man put down his buckets and set his hands to the stone. He couldn't read it the way that he could have read a tree but he walked his fingers across the surface until he found cracks large enough for his hand and began to haul himself up hand over hand. Before long his buckets were vanishingly small dots at the base of the cliff face, and a short eternity later he raised his hand and it met only thin air.

Green Man felt his stomach fall out through the soles of his feet as he flailed for a handhold, and then out of nowhere gnarled fingers caught his wrist, and he felt himself lifted bodily onto the summit of the mountain. In the distance, the sun was rising.

"You made it," said a very old man, with a face as worn as the mountainside and a beard like tumbled boulders.

"This is the office of La Fundación Rigoberta Menchu, how may I direct your call?" the voice on the other line spoke in lightly accented English.

"My name is Reginaldo, I'm calling from Minnesota, I would like to speak to…"

"They told me about you," the voice switched to Spanish, following Regi's lead, "I'm sorry, but she's busy right now."

"Look," Regi said, "I'm going to keep calling every day until you put me through to her. My time is probably worth less than yours right now so I won't hesitate to waste it."

There was silence at the other end of the line, then a sigh, "Let me put you on hold."

Regi punched the speakerphone button and tinned Guatemalan folk music filled his cubicle. It had been a strange road that led him here, and it all started with the cases of crafts that he hadn't been able to sell.

Once Regi figured out that white, rural Minnesota wasn't a good market for the goods he was selling, he started looking for other venues. With Amy's help he started a company called Guatemala en Vivo (Guatemala Alive) with the stated goal of helping Guatemala's indigenous people earn an independent living.

Everywhere that Regi sold crafts he also began to sell a message: there is no way for many Guatemalans to support themselves without buying into an exploitative system of migrant labor, but crafts can provide an alternative income and help break the cycle of poverty in the villages. The message, however, proved to be more popular than the crafts.

At the same time Rigoberta Menchú published her memoir, "I, Rigoberta Menchú," and her work as a labor organizer and peace activist rocketed into the consciousness of activist communities around the world. Regi found himself at the center of a growing awareness of the struggles of Guatemalan farmers, and while he still wasn't able to sell enough crafts to make a difference, he began making contact with all kinds of nonprofit groups that wanted to know how they could help.

Regi had met Niel Ritchie at a meeting of the Environmental Quality Board and something clicked right away. Niel was working on farming issues with the Institute for Agriculture and Trade Policy (IATP), and he saw an opportunity for Regi and him to help one another. It was 1994 and a major farm bill was moving through committee in Washington. IATP was interested in bringing small farmers south of the US border to the table to make sure that their needs were represented and to do that they needed a translator—someone who could speak Spanish, but also who could speak the language of small farmers, someone who understood the world they came from.

Regi took on the role of translator and built a strong relationship with the Union de Ejidos de la Selva. The Union represented coffee growers in southern Mexico, and by the time the farm bill passed Regi had established good relationships and was already thinking about ways to bring his experiences together.

He was still certain that trade was the way to give farmers ownership of their own destiny, but it was also clear that the trade in crafts simply wasn't enough. Now he had contacts in the coffee world, and that market was already well-established; it was just a problem of getting the coffee he wanted to sell.

Working directly with small coffee farmers in Guatemala is not just a matter of good business intentions. Coffee exports are controlled by a group made up of primarily large plantations which deliberately keep small growers out. Regi started by working with growers in the Chiapas region of Mexico. First he tried partnering with an existing company called Aztec Coffee, but the quality was poor and substantially more expensive. So that idea didn't work.

Regi had grown tired of drinking the burnt sludge that passed for coffee in most of the U.S. and had been watching in amazement the rapid growth of specialty or so-called "gourmet" coffee shops popping up on every corner every day. He saw an opportunity to introduce Minnesota to the sort of coffee that he had grown up with. Different from most coffee growers who export the best part of their crop and drink bad coffee themselves, Regi's father didn't care to export any of his crop and had planted coffee under the banana and avocados shade and they hand-processed and roasted their own, selecting the highest quality for their home and selling the rest to the neighbors. He arranged to import the green coffee beans from a cooperative in Chiapas and branded it Cafe de la Selva. Regi worked with a roaster in New Prague to develop specific roasting standards. At last Regi had a product that he could sell, but he still needed to find a way to source coffee from the struggling farmers that he dreamed of helping.

When Niel's brother Mark Ritchie approached him about a buying trip through the Guatemalan highlands with a wealthy New York businessman, Regi didn't waste time asking too many questions. He was willing to guide John through war-torn rural Guatemala if it would give him a chance to make contact with independent farmers.

John had read Rigoberta Menchu's book and wanted to see the places and the people that she wrote about, so when they had finished their rounds of the relatively safe markets in San Cristobal and Momostenango in the province of Totonicapan, Regi took him into the part of the country that had seen the worst of the fighting.

They stuck out like a sore thumb in the insular indigenous communities and John suffered severe gastrointestinal problems because he insisted on eating where the locals ate and drinking non-bottled water. Regi was constantly on edge, working hard to keep the native Guatemalans from growing uncomfortable in this stranger's presence. He would oftentimes ignore John's requests to translate and just carry on mundane conversations with the people they met. John didn't understand that people might not sell him the clothes off their back, or that inviting himself into a person's home meant they had to cook him a dinner with whatever they had, even if it was all they had. It didn't help that everyone knew that the United States had trained and armed the military death squads. General Gramajo, the head of the armed forces of Guatemala during the thick of the war when large indiscriminate massacres were carried out in these villages, is known to be among a large number of Latin American military leaders in the hall of fame at the army School of the Americas in Fort Benning Georgia, renamed the Western Hemisphere Institute for Security Cooperation.

By 1995 the military and the rebels mostly respected an informal ceasefire, but the war hadn't officially ended, and the region between the Ixil Triangle and the Pan-American Highway was still thick with soldiers. It was also home to some of the best markets in the country, so John insisted that they stop at every decent-sized village.

At first Regi did his best to humor the impetuous American, pointing out the rubble and collapsed cement block houses or gaping holes left by tankfire, or the places where corn grew unusually tall and green signalling the presence of a mass grave, but anytime he pointed something out, John wanted to investigate it.

When they stopped at the market, John would interrogate the sellers, asking where they had bought their crafts and how much they had paid. Regi translated, and explained to the frightened locals that John didn't know that what he was asking was potentially dangerous. He had tried to explain it to John while they were in the car together.

"You may not think twice about asking how much someone paid for their crafts, but every town has a military 'ear' or more than one, and if they hear that some gringo is walking around asking about prices, that's going to make them nervous. They'll think that you're trying to agitate for better wages for the artists."

"So?" John said, "I'm not, and even if I was, what harm could it do?"

"To you? Not much," Regi admitted, "and the military isn't likely to do anything while an American is watching. But after you leave?

Those people are vulnerable, and people have been disappeared for much less than answering awkward questions."

"That's ridiculous," John said.

Regi spotted a large boulder blocking the road ahead. It might have fallen naturally, but it might have been pushed there, and either way there was a good chance that there were men hiding in the trees ready to take advantage of unwary travelers. "When I was younger I took the chicken buses if I needed to travel anywhere. You've seen them?" He checked his speed—he was going fast enough that if he didn't slow down, no sharpshooter would be able to draw a bead on him. This maneuver wasn't going to make John happy, so Regi worked to keep him talking.

"Of course," John said.

"Well at every checkpoint the bus would stop, and the military would order everyone off. You could usually tell one was coming up when you saw smoke on the horizon."

"Smoke?" John said.

"From burning houses." Regi shifted, picking up speed. He had spotted a track off the road that led around the boulder. He was glad he wasn't the first one to come this way. "I went through probably hundreds of checkpoints, and at every one of them someone was left behind."

"There was always a rebel on the bus?" John asked.

"That's not what I said," Regi swerved off the road. He thought he could see people moving in the trees, he swung back onto the road and pushed the car as fast as it would go for the next mile. "I said we always left someone behind." He realized that John was holding onto the door, white knuckled.

"What the hell was that?" John said.

Regi shook his head, "They did it just to remind us that no one was safe."

"Jesus!" John said.

After that, Regi stopped pointing so many things out to John, and the rest of the trip went smoothly. But Regi couldn't help but wonder again and again, what would have happened if he hadn't taken the opportunity to leave his home when it came. What were the chances that it would have been his body in one of those shallow graves feeding the corn?

Think about the future, he told himself, *You can't shape the past.*

While they traveled they talked a lot, and John agreed to give Regi the seed money he needed to start importing fair-trade Guatemalan coffee. When he got back to Minnesota there was only one piece that Regi had to put in place, which was why he was spending his afternoons on hold with La Fundación Rigoberta Menchú.

There was a click and for a second Regi thought that the receptionist had hung up, "Are you still there?" came a voice on the other end of the line.

"I am," Regi said.

"Alright," the receptionist said, "make it quick."

"Thank you," Regi said, and Rigoberta Menchú came on the line. In the United States people liked to think of Rigoberta Menchú as a simple village woman, but she is actually whip-smart. Regi outlined his plan to import coffee directly from the growers to help support more small coffee growers, and shift power away from the cartels. Rigoberta Menchú cut to the chase.

"So what do you want from me?" she asked.

"I'm looking for your endorsement," Regi said.

"Alright, I've got two conditions..."

To carry out his trade work at IATP, Regi had incorporated a holding company under the name of Headwaters International, owned 100% by IATP. After a few weeks and a full endorsement of the idea by IATP, Regi had a product called Guatemalan Peace Coffee to get started, with an endorsement from a Nobel Prize winner. Three percent of the Peace Coffee profits would go to Rigoberta Menchú's foundation.

The idea took off, and soon they were selling enough coffee that Regi had to bring on more staff. As the company grew, so did the need for more capital. Using the small equity they had accumulated on their newly purchased home, Amy and Regi refinanced it to take a loan to buy the first container load of Guatemalan coffee. The growth strategy included the consolidation of Cafe De la Selva and Guatemalan Peace Coffee into one brand. The idea of Peace Coffee was a simple continuation of the original vision, but also something that could be applied to most of coffee-producing countries. He bought the domain name on the Internet, reserved a toll-free number, and set in motion the hiring process and office organization needed to bring in someone to continue the consolidation and growth. As soon as the new person was hired, Regi began to step back—he knew when it was time to get out of the way and let people with more training and knowledge of the U.S. culture and markets take over.

Regi had a knack for starting companies and was self-employed in a variety of ventures, each distracting him from finishing his college degree, although all fulfilling in their own way. Still, living in Minneapolis was hard for him. He planted the backyard with black beans and vegetable gardens to feel more connected to the land, but it was a far cry from farming. The family had grown with the birth of their daughter, Nicktae, and the arrival of Regi's young brother Adolfo who had come to study in the U.S. Nicktae was now a toddler and it was already clear that she was her father's nature girl. Life in the city was not what Regi had envisioned for his children, yet unintentionally they were becoming city dwellers.

It was the year 2000. Willito was about to begin school and Regi wanted him to learn beyond the walls of a classroom. Amy and Regi decided to move back to Guatemala for a year to immerse their children in the language and culture of their ancestors. They yearned for an opportunity for the kids to connect with Regi's family and the land. Regi's younger siblings were now establishing their career paths, Carmen as a teacher, Cesar pursuing carpentry, and Elias a study of tourism. Regi wanted to be closer to support them as he has been supported by his older siblings. One brother in particular, René, now graduated from agriculture school wanted to start a sustainable forestry business.

So Amy and Regi planted trees and helped to establish an area of forest near a house—a forestry office building—that they built near the orphanage where they had met. Even though the war had come to an end with the signing of the peace agreements in 1996, the country was still violent. It was more dangerous than Amy and Regi were comfortable with as an environment to raise the children. After a year, they returned to Minnesota and began looking for ways to create a farm of their own.

Now that Peace Coffee had taken off and Regi had moved on from that endeavor, he was able to dedicate himself to pursuing a degree while working seasonally as they figured out their next move. It was clear that Regi could not keep living in the city. Farming was in his blood.

They decided they would work to find a farm and move as soon as Regi graduated. The question remained, "How?!" With Amy earning a teacher's salary and Regi working part-time in order to finish school, the savings account would not support a down payment for even a small piece of land. Further complicating life, William had developed a mysterious high-fever illness since their return from Guatemala which meant medical bills were piling up and Amy needed to step down from her job to create an income through running her own preschool in the upstairs of their duplex so she could be home with William. Slowly, William recovered, just as life threw another surprise their way. They were expecting again. Lars Decarlo came into the world as a happy healthy baby. Decarlo they called him for the first few days. "Of farmer," they willed the name to bring what now seemed impossible. The farming dream seemed unattainable. That was until a chance meeting with some friends of friends.

Hans and Heidi, owned a farm outside of Jordan, Minnesota, and were interested in transitioning to a more sustainable model of agriculture that included a community farm. Heidi owned the land, and Regi worried that may cause conflict if he and Amy didn't establish a level of ownership themselves, so they wasted no time moving into a small trailer on the property.

Regi spent the first year commuting to the Twin Cities where he earned a living building decks and fences while they worked on a house on the farm. Without buying land development rights from the neighbors, Amy and Regi couldn't get a permit for new home on the property, but there was a small 1870s log building already on the property and they began to rebuild it with the hope that it would serve as an interim home until they could purchase some additional land of their own or work on a different arrangement to get the building rights on the land.

Regi began growing food and together he and Amy did a lot of dreaming on how the farm could reflect the shared vision with Hans and Heidi's work. He took his memories of the farm his family had kept, with its layers of growth, and strategic intercropping, and mapped out ideas for adapting the process to fit Minnesota's climate and growing season and the ecology of the farm. The model closely paralleled the permaculture system and so they embraced the name that was developed for the farm Seven Stories—after the "seven layers" described by the agricultural theorist Bill Mollison.

In the first planting season they put in eleven acres of trees, experimented with different strains of hazelnuts, and established a good mix of food crops. Regi relished the work, and finally having the room that he was used to. He approached two neighbors to the east who owned woodlands and got permission to hunt there, he got back into practice target shooting with a rifle and bow.

It wasn't until winter that the trouble started.

Not long after the first snowfall, Regi started noticing snowmobile racing across his fields. When he approached their neighbor, he shook his head.

"Look," he said, "I like you. I don't mind having you around, but that's not how everybody feels. There were some people around who were surprised when you didn't leave with the snow. They came to me asking what was going on."

"What did you tell them?" Regi said.

"I told them you were a part owner in the farm. They didn't take that too well."

"So they're driving across my fields?" Regi said, "ruining our trees?"

"Relax," the neighbor said. "It's probably just kids being kids. They can't tell where that access road is with fresh snow on the ground. I told you about the neighbors just as a heads up, I don't expect anybody will try anything stupid."

"I hope so," Regi said, "but you tell them from me that we don't have a public access road on the property—that's an access easement for the electrical company not public property and they need to stay off it."

"I'll see what I can do," the neighbor said, but he didn't seem hopeful.

The snowmobilers kept coming, and eventually Regi stopped trying to track how much damage they were causing. He began to have bad dreams again, dreams he hadn't had often since leaving Guatemala. He dreamt that he and his family were surrounded, that shadowy figures were coming for them and there was nothing he could do. Many nights he woke up screaming, covered in sweat.

Target shooting calmed him down. Each time the dreams came back, he set the target further away and settled in until he could hit the bullseye reliably. The repetition helped him put the dreams to rest for a little while, but it never worked for long. He felt afraid; he had already studied cases in other rural areas where immigrants had been hurt, asked to leave at gunpoint when they showed up to collect their paycheck after weeks of work on the farms. In other cases, farmers and meat processors especially were known to call immigration to have undocumented immigrants deported before payday. Regi was already feeling on the edge. He had tried his best to fit, laid low in town and at church, which was difficult for someone used to running things and shaking things up. But he knew the culture was not on his side; all he hoped for was that he could just farm, as for the rest he did not mind being invisible.

The next planting didn't go well. When the snow thawed, people started riding dirt bikes and four wheelers through the black beans field. Regi went to the local sheriff but he was told that until someone got hurt or the police came while the intruders were on the property, there was nothing the police could do.

Regi started keeping the keys in the ignition of the truck while he worked. Twice while working on an access road by the main highway, guys on trucks drove into the ditch between him and the highway and came right at him turning away just feet away from him. He never moved, in the moment he never felt scared, even though a simple mistake of the driver and the truck would have ran him over. Once the occupants of the truck were hunters and he could see the gun's through the back window of the truck when it drove away. The riders were coming through during the daylight now and it was clear that they wanted to provoke him to do something; he had heard of plenty about "accidents" of this kind.

Things came to a head one afternoon when Regi and his younger brother Adolfo were working in the bean fields. A biker passed so close to him while in the field that he had to move to not get hit. Regi bolted for his truck and angled toward the easement to cut off the rider. He couldn't make out much about the figure—he was wearing heavy clothes and a tinted helmet and riding a mostly new dirt bike.

There was an alfalfa field at the end of the easement road and Regi swerved the truck across the first few rows and pulled the hand brake cutting the biker off so he had to stop to avoid hitting the truck.

The biker skidded to a stop at the last second, Regi looked at him straight in the eye leaving no doubt he was ready for whatever needed to happen.

"Take off your helmet," he said.

The biker didn't move.

"I need to know who I'm dealing with," Regi said, "take off the helmet."

Moving slowly, the biker unbuckled his helmet and dropped it in the dirt, "Take it easy man," he said.

Regi shook his head—the biker was a boy, round-faced, he looked scared. In Guatemala if someone was trying to run him off his land, they would be a grown-up. They would be serious, and they would be armed. This was just some kid playing games, although it was clear that it was directed games, so there were grown-ups involved—that Regi knew was for sure, he knew enough to guess the rest.

"What's your name?" Regi said.

The boy stammered a reply.

"Get off my property and tell your friends to stay away too. I don't ever want to see you again."

"Of course, of course," the boy said, "I'm sorry." He bent to pick up his helmet and almost fell.

"And take the easement back," Regi said.

The kid nodded as he crammed the helmet back on his head. He wasted no time getting away. Regi swallowed, he knew better than to think that this was the end of the issue.

A few days later he was working on the roof of the log cabin. It overlooked the driveway that led up to the main house and every time he heard a car coming Regi found himself looking up in fear. He felt terribly exposed on the roof and on top of it he was wearing a harness tied to a rope, unarmed and without cover. After a few hours a strange truck pulled into the driveway and Regi felt his blood rush, his heartbeat accelerate, his legs heavy. In his head, he started to run possible scenarios.

He breathed deeply, and willed his muscles to loosen up as the truck pulled to a stop and a man leaned out the driver's side window.

"Hey," the man called.

"Hey," Regi said. He squinted, trying to see if there was a passenger in the truck, but there was too much glare off the windshield.

The man looked around as if making sure no one else was around, but Hans had come to help Regi with the roof and he was on the opposite side, hidden by the roof ridge

"You hunt the land behind this property?"

"I have permission," Regi said.

"Sure. It's just, I've got a friend who hunts back there too—he's got permission see, and he's pretty territorial. He comes across you by surprise, he might shoot first, ask questions later, get me?" the man said.

"Your friend have a name?" Regi said. "Maybe you can have him come talk to me himself."

"No need to get hostile," the stranger said, "I'm just giving you a friendly warning." There was nothing friendly in his tone.

"When does your friend usually hunt then?" Regi said.

"He goes out after church most Sundays," the man said.

"Alright," Regi licked his lips, "you tell your friend I'll be in the woods this Sunday, three in the afternoon. He can deliver the message to me then."

The stranger shook his head, "Your funeral," he said, and pulled back onto the road.

On Sunday Regi went into the woods as promised. He waited in the open with his rifle pointed at the ground for a full hour. The birds sang brightly, the day smelled of pine and damp, and the sun painted the ground in pools of light and shadow. When no one appeared to meet him, Regi laid on the ground face up; he meditated on the sounds, tried to guess what made them, focused his hearing. This place was a world of its own, beautiful, spiritual, he meditated on the idea that someone, the truck man for sure, wanted to hurt him and of all places, he wanted to do so in a place like this. Somehow he felt much safer in the forest. Then he walked home.

As he passed the newly planted fields, his thoughts drifted to the past few months. The problems with the neighbors were only a piece of the growing difficulties with farming here. The county officials were being called many times checking up on different reports of "illegal" activity happening on the farm. Once the inspector inquired about the "migrant" family living on the land and the legal status of all involved. Heidi and Hans were nervous that the farm would never get the needed variances to establish itself as a community farm. External tensions were directly causing internal tensions as well.

He took a longer time to walk back to the cabin, the fields were bare of whatever promise they used to hold for him. He felt hollow and too aware of his skin. That night he told Amy all that had happened, and they started making plans to leave. "Maybe that means they win," Regi said, "maybe I don't care anymore."

It was a natural choice to look for a new home near Northfield. It was where Regi had first lived when they moved to the United States, where Amy had gone to college and one place outside of the city where they had felt home before. The land they found was a short drive from Northfield, with its strong schools and food coop, and it finally seemed to Regi that he had the space and the freedom to start over again, this time things had to be different.

The tiny 1.9 acre farm came with a finished house. The house was also small, just two bedrooms, one-and-a-half bathrooms, but a mansion compared to their previous living arrangements. Amy and Regi sat in the backyard and watched the sun set over their neighbor's fields. "What do you think?" Amy asked.

"I think.... I think I'm glad Seven Stories didn't work out," Regi said. He thought about the suitcases that he brought from Guatemala, and the folder full of plans for the farm that he had dreamed of, "This way it's all ours," he said.

Regi had a lot of work ahead of him. The previous owners had trashed the place, and before he could start any work in the fields he spent hours walking back and forth across the property collecting tin cans, bent nails, and other debris. He dug a car engine out of the floor in the collapsing barn and filled recycling bins with old beer bottles.

The soil was parched and thin, but there were fewer rocks than the fields he had grown up with, and he began the long process of rebuilding the soil with a smile. The long hours in the field felt like coming home.

Not long after they arrived in Northfield, Regi heard about a community service position that was opening with the school system. One of the Northfield elementary schools had a huge Hispanic population, mostly the children of poor farmers, migrant laborers and tenant farmers.

Regi applied for the job, pitching a community garden as part of his commitment to supporting the children and their families. He thought back to the program that he had started to develop at the Catholic mission so many years ago in Guatemala. He would bring the community together to grow and eat food and help the teachers bring the lessons out of the classroom, so much math, biology, science, chemistry, visual learning, art—so much a garden can teach.

He got the job and threw himself into his work. He met with the parents of students and got a sense of what they needed. Just as he had always done, Regi started to imagine ways to meet those needs. What he envisioned went beyond a simple outreach program, it was a system, a system like the one he had worked on at the agriculture school. It would be a self-sustaining agricultural ecosystem, an ecosystem that would regenerate itself, that would contribute to the earth rather than stealing from it. The idea of the community garden became a reality as planned, but by then, Regi had already moved his thinking to a new vision he knew was possible.

By the time he stepped down from the position, his small farm had begun to take shape and Regi was able to reach out to universities and members of the community. He proposed experimental farming practices and won initial support for his ideas. Regi had not talked to Niel Ritchie more than a couple of times since his time at IATP, but he found his number quickly, called him and they met at Maria's Cafe on Franklin Avenue in Minneapolis. Regi proposed an idea of working on agriculture; Niel invited him to join Main Street Project, an organization he had founded just two years before. Regi brought his thinking and the projects he had started under the organization and work toward Main Street Project's Regenerative Agriculture Program began in the summer of 2007. It took several years for the project to take root, but finally his work and his dream was all coming together.

With the local programs established, Regi had begun to reach out internationally. The local programs were designed according to a larger idea Regi had incubated since the agriculture school days, the same vision Regi had shared on those long conversations with Amy about a future farm, a place where people would come to learn a new way of agriculture. What Regi was designing was not something for Minnesota or for Northfield, he envisioned much more. This included a network of internationally connected farm training sites, and Guatemala was back on the top again. After all, it is where it had all begun.

After having started the forestry business, Regi's brother René was married to Finola, a Canadian citizen. Soon after their marriage, gang violence drove them out of the area and they moved up to Canada, leaving behind the operation. Since then, Regi had been looking for an opportunity to use the resources that were left behind. Arge was back working at the orphanage supporting the new sisters-in-charge. While Regi's youngest brothers, Cesar and Elias, wanted to use the house for a hostel, Regi and Arge concocted a new plan to use the land to grow food for the orphanage while teaching the kids about organic and agroforestry systems. Things were looking up and the dreaming machine was back on.

And then the house in Northfield burned down. It was a quiet evening, the night before Thanksgiving 2011. William and Nicktae were both spending the night away and Amy and Regi were sitting in bed reading; outside they heard a train rumbling by.

It was a noise so common that neither one of them noticed anything unusual for a very long time. But the noise kept going. "That's an awfully long train," Amy said at last.

"It is…" Regi said. Amy got out of bed and padded to the door, the knob was cool but she could smell something sharp and foreign. She pushed open the bedroom door.

The hallway outside was awash in flickering orange and red light, and flames were crawling up the far wall of the house. Smoke was already pouring in under the door to Lars' room at the end of the hall and Regi immediately clicked into crisis mode.

"Get out of the house!" Regi said as he grabbed his backpack and cell phone automatically and slipped barefoot into his work boots. "I'll get Lars and meet you out front." They both dashed down the hall, smoke was pouring in under Lars' bed, but the fire alarms hadn't triggered yet. Luckily Lars was the only child home that night.

Lars was large for an eight-year-old, but Regi gathered him in his arms like he was weightless and took off running toward the door on the south side of the house. By the time they reached the foot of the stairs the house was full of smoke and Regi struggled to see. The front entrance and the deck above it were completely on fire.

In the yard they finally heard the fire alarms go off and Lars began crying. Regi set him on the grass next to Amy feeling strangely detached from everything.

Regi fished for his cell phone, dialed 911, and handed it to Amy. "Get firefighters out here," he said. "I'm going back in for the dogs."

"Don't!" Amy shouted at him. "It's not worth it!" But Regi was already moving.

The fire had spread across the front door so Regi headed back to the south side where a greenhouse attached to the outside of the wall led into the house. In his head he was mapping the quickest route to the basement, the dogs, and the things he needed—leashes, he would have to grab leashes. He heard Amy's voice as though it was coming from a very long way away and didn't quite register what she was saying.

Regi was opening the greenhouse door when the house exploded. Regi was thrown back, landing in the dirt. He lay, winded, and watched the flames race down the halls of his home. He knew there was no going back for the dogs anymore.

As Regi picked himself up he found that he was no longer scared. It was like a cold wave washed over him. Just about everything they owned was gone or would be soon, and he found himself thinking what could he save.

The fire had jumped to a great hollow silver maple tree in front of the house that overshadowed their truck and the family car. Regi ran to the truck and drove it out of the driveway to the other side of the street, then he went back for the car.

The vehicles safe, Regi gathered Amy and Lars and guided them across the brush and onto the gravel road to the south of the house. Regi walked back toward the house and turned off the line to the gas tank. Then he walked back to the road and waited for the fire truck to arrive. Regi tried to fight the intense feeling that he was missing something. It wasn't long before a state trooper arrived on the scene, and they huddled together in the back of his squad car until the paramedics arrived. Gradually Regi relaxed as it became clear that they were in good hands.

By the time the firefighters pulled up the house was fully ablaze, and they immediately got to work on containing the fire. They sprayed down the outer walls as best they could and saturated a circle of ground and nearby trees with water.

Regi asked if there was anything that he could do to help and the fire chief shook his head, "This is our job. Just get some rest. You got somewhere to stay?"

"My parents live in town," Amy said. "They are traveling out east; we have a key to their house."

"Good," the chief said, "go there, rest. Deal with this in the morning."

By the time they drove to Amy's parents' house, Regi was exhausted. Nicktae and William arrived and the family soon fell asleep, the events of the evening too unreal to process. It was only in the morning that Regi realized how little he had carried with him as he stood in the bathroom and wondered how he was going to brush his teeth. Then he realized that he still didn't have any socks or underwear. He had left the house without even a shirt on his back. He sat down on the edge of the bathtub and wondered how they would ever recover from this latest setback.

By eight in the morning their pastor was at the door with a box of five prayer shawls, and after him came a steady stream of friends bearing meals, clothes and promises of support. Over the next week they found a house to rent and moved in. News of what had happened began to spread through the community, and the donations kept coming until Regi had more clothes than he knew what to do with.

Friends of William and Nicktae took to Facebook to organize a community benefit for the family and within a week it seemed like anyone from their school who could play an instrument or carry a tune had signed up to help. High schoolers carpooled to the Northfield Ballroom after class to help decorate for the benefit concert. By six o'clock on December 16th, almost a thousand people packed the room and Regi could only watch in wonder.

"What did we do," he said under his voice, "to deserve this?"

At intermission and after the show people kept coming up to Amy and Regi and thanking them for things Regi couldn't even remember doing; favors as simple as driving someone to the airport, or having them over for dinner. "That's different, anyone might have done that," Regi found himself saying to one of Amy's oldest friends.

"It doesn't matter," she said. "It's what we needed. This is what you need, so this is what we can do for you. I hope we're never in a position that you can return this particular favor."

"Still, thank you."

"That's what community is for," she said.

After the concert there was a bake sale, and still more donations came in. The support was overwhelming. Once the rubble was cleared away, so many people turned out to help clean and restore the foundation that they got the whole job done in one afternoon.

Regi was amazed to find that it didn't even feel like starting over, a feeling he no longer appreciated—instead it felt like moving forward.

The house was finally hooked up to the grid and move-in ready midsummer of 2012. It was the work of many people's hands, old friends and new, all contributing as much as they could. On a warm August evening they held a dedication ceremony with friends and family from as far away as Guatemala. The house was blessed with traditional Guatemalan prayers and dancing and the dinner table was heavy with donated hot dishes, roasted chicken, tortillas and black beans.

*T**he fading twilight was filled with laughter and songs and that night Green Man walked through the cornfields under strange stars.*

"I did," Green Man said. "Now what?"

"Now the real work begins," the mountain replied, "if you want to save your home."

"I do," Green Man said, "but I don't know how."

"Everything is connected Green Man," the mountain said. "I can feel the oldest betrayals and crimes against the earth shivering up through my roots. Decisions etched in the stone of ages are passed on to the people of this moment and to their children. That is what you face, a legacy of poor choices."

"What do you mean?" Green Man said.

"The rain that fed your forest traveled on the wind from the sea for time out of time, but the winds have changed. I do not know why. The wind reaches me with barely a whisper now…"

"So what can I do?" Green Man said.

"You? You're Green Man—you can run down the wind and ask her why she doesn't blow so strong. I called you up here because of all the creatures in this world, you are the one that never gives up. If you can't do the impossible then no one can." The mountain sighed, and the stones trembled beneath Green Man's feet.

"I'll do what I can," Green Man said, "but I don't know how to find the wind."

"Look," the mountain said, and held out his hands. They were cracked and gray, and as the thin wind whispered across them his hands dissolved into fine dust. Soon Green Man found the dust all around him as the old man wore away to nothing. It swirled like a ribbon and flowed up and was gone.

"I don't understand," Green Man said.

"Everything is a cycle," the mountain spoke with his all encompassing voice. "Watch the sea."

Green Man looked out over the ocean, and at last he saw a flash of gray and silver touch down far across the water. "How can I get there?" he said.

"You must find a way," the mountain replied.

So Green Man started down the mountainside and walked through his dying forest. A few animals watched him go, or followed him for a time, but when they spoke all Green Man could do was shake his head and keep going.

At last he reached the sea, and dipping his hands into the water he found that he couldn't feel a thing beyond the warm brine. He walked up the beach and pulled down several young trees that he lashed together with rope. Above him he saw Vulture circling, but she didn't come down to talk to him. At last, Green Man pushed out to sea.

He caught the retreating tide and rode it out onto a still ocean. Without a strong wind the water was glassy and Green Man lay on his raft and paddled with his hands. He moved achingly slowly and the sun roasted his back like tree nuts in a fire. Whenever he worried that he had lost his way he would look up and see the mountain dust circling, an ever-narrowing thread leading him on.

That night something moved in the deep and Green Man drew his arms and legs onto the raft and hoped that it would not see him. And so he slept.

In the morning, by chance or some plan, his little boat ran aground on an alien island dusted with mountain earth. Green Man stood on the beach awhile and looked around. The island was small, but covered in shipwrecks, in the flotsam and jetsam of ages past. There were animals too, lost-looking turtles and the strange serpents that made their homes in the deep.

He didn't notice the tide pluck his raft up and wash it out to sea.

"Hello?" Green Man called, picking his way through the debris on the beach, "I seek the wind."

"Of course you do," a voice whispered back. "Why do you think I brought you here? The mountain told me to receive you."

"If the mountain can speak to you, why did he send me?"

"Because the mountain is, and always will be. You are young and hurting, the child of men, child of earth," the wind laughed quietly. "He thought that I might pity you."

"Why don't you blow so strongly anymore?" Green Man said. "Why don't you carry water to the forests?"

"I am choked," the wind said, "by the smoke of many fires. I will not blow where I am not wanted."

"What do you mean?"

"I mean that the humans do their best to still me. Why should I carry water for them? Or for you, my poor earth-child?"

"Without you, trees and animals are dying. We need you more than the people in their cities could possibly know," Green Man said. "Come back for us."

"Everything is connected," the wind said. "The people in their cities eat food from your forests. Besides, it is too hard to gust weighed down as I am with garbage."

"What can I do," Green Man asked, "to make you blow again?"

"Nothing," the wind said. "To visit the shore is to make myself unclean with the smoke and soot of factories and I will have no more of it.

"What if I could clean you?" Green Man said. "I know the trees and creatures of the forest. The monkeys would pick the ash from you just like they pick insects off of each other."

"Too little," the wind said, "and too late."

"Let me try," Green Man insisted, "or else I will die on your island and clutter it even further."

"Very well," the wind said, "I will carry you to shore with as much dust as I can carry. If your forest can scrub me clean, then I will carry you water once again.

And so she did. When Green Man landed on the beach, he ran to the rainforest, stretching his wits out ahead of him, feeling for every root, leaf and flower that would answer to his will. Bit by bit he wove it all into a net.

When the wind passed through the rainforest, the dust and dirt and grime caught on the leaves and dropped to the ground. Green Man heard a faint, "Thank you," as the wind lifted up, free from her burdens.

It wasn't too much longer until rain came back to the forest, and the trees drank in life, and the animals played once more. But Green Man never took his world for granted ever again; the impermanence of it all frightened and thrilled him.

Chapter 13

Green Man's Daughter

"I think," Regi says, "that there is only one story left to tell."

I realized that I am stiff, that my hand has cramped from writing. I don't know what time it is. "What's that?" I say.

"What happens next," he says.

"What do you mean?"

"I mean that even Green Man couldn't change the world all by himself, and so one day, while he was walking through the woods, he heard a baby crying…"

"Ah" I reply with sudden understanding, "the next generation."

"Yes," Regi nods. "It's really all about the kids. All the thinking I've done since becoming a parent has involved my children."

"But none of your children are aspiring to be farmers, are they?"

"Not necessarily, but they have each taken the experiences that we have learned in this journey together and I know their work in the world will be about giving life. William, he's an actor, a performer, not too keen on working in the field right now, but he has learned the value of working hard. He took on raising a flock of chickens on his own as a teenager and has helped me with research on the birds. He is interested in the way the world works and biology; it would not surprise me if he ends up back in a garden someday."

"What about Nicktae and Lars?"

"They've had more time in farming. This is the only home that Lars remembers. He doesn't know life without chickens and a big garden. Nicktae too, just as a little girl in Jordan, she embraced farm life fully. She loved our simple log cabin without electricity and living off the land. When we got to Northfield she jumped right into chicken care, even raised a couple goats for a while."

"Goats?" I asked, raising an eyebrow.

"Yeah, that was a mistake I'd rather not talk about," Regi laughed.

"But, isn't Nicktae studying art now?"

"Yes, but she's stayed connected here at the farm. She is part of the YPAC group."

"YPAC?"

"Young People's Action Coalition. The next generation isn't just about my own children. This group of young people have started a farming cooperative they have named Finca Mirasol."

"After your farm?"

"Yes, I started training them on our own space and they started using the name. They can have it if they find value in it. Nicktae has been involved farming as part of that group growing black beans and onions and doing a lot of study of the food system. They are an amazing bunch of youth. They inspire all of us."

"So she's your farmer?"

Regi laughed, "Some days at least!"

"And Lars?" I asked.

Regi chuckled again. "Our Decarlo. We have come to call him our farmer boy. Of the three, I think he's the most likely to go into farming. He has been interested in farm projects from a very young age. He started a business producing and selling eggs as a fourth-grader and he gets excited when we talk about farming. But he's still young. Who knows what direction life will take him!"

"You seemed to know from a very young age." I pointed out.

"Yes, well I had Green Man to guide me."

"Don't your children have him as well?"

"I suppose they do," Regi smiled. "I suppose they do."

It was the height of the dry season and the heat was enough to make a person wilt, even in the shade. Green Man's garden was growing well, and it had been a long time since he had seen soldiers in his part of the forest, and so he walked without real purpose just checking in with the earth and the animals in the trees. He lay for a while in the cradling roots of one of the great trees and listened to the wind rustling the leaves high above his head.

He found a crack in the ground where clear water bubbled through and he washed his hands and face there and relished the cool relief on his skin. A troop of monkeys found Green Man and they held their noses and laughed at him.

"You stink Green Man," they said. "You should wash more, wash all over!"

Green Man made a great show of smelling himself and then shook his head, "I'm sorry," he said, "if I smell, I didn't know. Your stink has clogged my nose for the last two kilometers."

Then the monkeys threw dirt at him and scampered into the trees, and Green Man chased them. His hands found cracks in the tree bark that even the monkeys missed, and he relished the feeling of flinging himself through the air. When he caught the monkeys, they wrestled, and when they were tired they fled into the canopy and he let them go.

"Tired old Green Man, stiff as a tree," they chanted as they fled. "Too old, too slow, can't catch me!"

Green Man just chuckled to himself. He splashed water across his face and the back of his neck. That's when he heard the crying start.

Because Green Man lived so far from other humans at first he wasn't sure what the sound was. He knew it wasn't a bird or a monkey or one of the predator animals. At first he thought it might be a young donkey that had been injured, or maybe some lost creature from far away, but when no parent responded Green Man decided to investigate.

The baby lay on a banana leaf as naked as the day, and Green Man stopped short of going to her because he knew that humans, like all animals, are protective of their young. As the crying went on, Green Man knelt and spread his hand on the earth. He pulsed warmth and good feelings to the little girl, and her crying faded. He could tell that she was terribly hungry.

Stretching out his senses Green Man tried to find the girl's parents, but they were nowhere nearby and he wondered that anyone would abandon such a precious creature when she was at her most vulnerable stage. It was not the first time humans had confused Green Man, and he was sure it wouldn't be the last.

He wrapped the leaf into a makeshift diaper, and carried the child back to his clearing. He started by feeding her mashed-up banana, but it was clear that wouldn't do and so Green Man went and stole a cow from pasture that gave good milk. He fed the girl by dipping his finger in the milk and letting her suck it clean.

Whether it was the milk or something about Green Man himself, the babe grew stronger every day. Soon she was walking on her own and causing all sorts of mischief with the animals in the area. Sometimes she would wander off while Green Man was tending the garden, and then he would have to chase her down, which left him less time to garden.

There were days that Green Man felt he was barely holding himself together, days when he wondered why he was being allowed to raise this child. And still she grew stronger, and still it took more work to keep her safe.

It was only when she started talking that Green Man realized what she was after.

"What's this?" she would ask, pointing to the leafy sprouts of one of the potato plants.

"It's a potato," Green Man said.

"It doesn't look like a potato."

"Well, the part that you eat is underground," Green Man said.

"Then why is this part here?"

"Because it helps the plant grow," Green Man said.

"But why?"

"Well, all life needs nutrients to grow, and the leaves convert sunlight into nutrients for the plant."

"But why?"

"So that the potato can grow, and when the plant dies a new one will sprout."

"That seems pointless."

"It's the cycle, Green Man said. "All life is a cycle. We grow and take energy from the earth, we die and return that energy to the ground and other life springs up."

"But why?" she said.

Green Man didn't have a good answer for that, instead he took his daughter by the hand and led her to his rows of black beans. "Do you know what these are?" he asked.

She thought for a minute, her brow furrowed, "Potatoes?" she said at last.

"These are black beans," Green Man said.

"But they're green!"

"Feel the pods here, each one of these little lumps is a bean just like we have for dinner," Green Man said.

"Does the plant grow them just for us?"

"No, they're seeds," Green Man said, "for new plants."

"But then we're hurting the plant by eating the seeds," she said.

"But we also help to care for the plant and make it grow well, and we plant new seeds each year so that the cycle can continue."

"I don't want to hurt anything I eat," the girl said.

"That's not possible," Green Man said.

"I'll just eat salt and sugar and lard," the child said, "because they aren't growing things."

"I make our lard from wild pigs," Green Man said.

"Then I'll just eat salt and sugar."

"Sugar comes from cane plants."

"Then I'll just eat salt."

"Okay," Green Man said. Her resolution lasted one spoonful into dinner and then the girl started crying.

"Shhh," Green Man said, "it's okay. I'll tell you what, I'll take you into the garden with me and you can see the way the plants grow."

"Okay..." the girl said, she seemed skeptical.

"Lets clean up now," Green Man said. "I bet I can wash my plate faster than you!"

"Oh, no, you can't!" the child said. Green Man let her win, the way he always did. That night she fell asleep easily.

In the morning Green Man took her hand as they walked down the path toward his garden. They walked by the low area where he planted his corn in the early season. It was flooded now and the muddy water rippled with hidden motion. Green Man kept the child close so that she didn't run off and provoke any snakes.

"Here we are," he said, when they reached the entrance to the garden.

"It looks just like the rest of the forest," the child said, disappointed. She stared at the rings of banana trees.

"It does from here," Green Man said. "Come on." He led her into the garden.

The paths branched and circled through the trees. "See there," Green Man said, "the coffee bushes grow underneath the bananas. The banana trees protect the coffee, and the bushes help to nurture the trees."

"But we don't eat much of either of those things," the child said.

"They help the rest of the food plants grow," Green Man said. "Do you see that?"

"The vine-y thing?" the girl said. "It looks like..." She broke away to look more closely, "Is it beans?"

Green Man nodded, "The banana trees provide something for the beans to cling to. If you look closely you can find them climbing the coffee plants too."

"Isn't that bad for the other plants?" the child asked.

"What does it look like?" Green Man said. Everything was green and heavy with fruit.

"It looks okay, I guess."

"Look underneath the coffee bushes," Green Man said.

The child looked at him, "Why?" She was suddenly suspicious.

"What do you think you'll find?"

"I don't know, bugs and stuff."

"I think you'll be surprised," Green Man said.

The child knelt and pushed aside the low branches of one of the plants. Green Man watched her face; her mouth was set, determined to be disappointed. He watched her break into a smile.

"There's so much!" she said. And there were potatoes, garlic, ropey strands of tomato. "How does it grow without sunlight?"

"It gets what it needs from the soil," Green Man said. "The larger plants take nutrients from the sun and return it to the earth. It's why the garden looks like a part of the forest, because it's structured the way the forest is—in layers. The bananas form the canopy, the coffee bushes are undergrowth. It's more than just a garden, it's a habitat for animals and bugs, and all of that life helps to nurture the things that we eat."

"But how?" the child said. "Why would they help us?"

Green Man reached out with his senses, looking for the rapid heartbeat of a chicken. He found one pecking under another bush, invisible from where they stood. "Look under that bush," Green Man said, pointing.

"Why?" she said.

"It might help answer your question," Green Man smiled to himself. "You might be surprised how often you can answer your own questions just by looking under the right things."

"Or you could just tell me," the child said.

"Yes, but then you would have to take my word for it," Green Man said. "This way you will know."

"Oh," the child said, "Okay!" She knelt down by the bush and lifted the branches with a stick. She poked the chicken on accident and it startled, flapping and jumping at the girl's face. She let out a shriek and fell backwards, and then started laughing when she realized what was going on.

"There was a chicken under there!" she said.

"It isn't the only one," Green Man said.

"But why?"

"Chickens are simple—what do they want?" Green Man said.

The girl thought a moment, "Something to eat?"

"Exactly."

"And the dark and plants and stuff make it a good place for the bugs that chicken like to eat?" the girl said.

Green Man nodded, "That's one of the ways they help us, by eating the bugs that might try and eat our food. I think of them as the cleaning layer of the garden."

"That's silly," the child corrected him. "They move around too much to be a layer."

"Okay," Green Man said, "you got me. But can you think of the other way that the chickens help us?"

"Ummm…" the child said, "jaguars eat them so they aren't hungry enough to eat us."

Green Man laughed, "I guess that might happen, but think what else the chickens do all by themselves."

The girl said nothing, her brow furrowed.

"What do you do after you eat?"

"They poop?" she said.

"That's right," Green Man said, "and that way everything that they eat is returned to the soil to help nurture our food."

"We eat chicken poop?!" the girl said.

"Well, not exactly..."

"That's gross." she crossed her arms.

"That's life," Green Man laughed. "That's the cycle."

The girl thought about that for a while. "Can I be a part of the cycle?" she said at last.

"Of course," Green Man said.

After that the child began to help in the garden, in little ways at first. She helped carry seeds and shuck beans. She stripped the leaves off of sugarcane and helped Green Man feed it into the hand-cranked mill. When she was old enough, she began to turn the mill wheel herself. She learned how far to space seeds from one another and which plants helped each other grow.

Over time the girl learned to read the patterns of the forest, its plants and creatures, and Green Man learned not to fear for her safety when she went walking alone. When she was ready, she helped him cut and burn a space behind the roundhouse where the ground was high and the light was good, and there she planted her own garden. Nearby she built her own house.

Green Man wondered at her, at how quickly she had grown and how she moved in time with the trees. When he looked at her in just the right light, he could even convince himself that she was a little, just a tiny bit, green.

"So you're saying there's a bit of Green Man in all of us?" I say. The sun is setting, casting red-gold rays through the windows and I wonder where the time went.

"Of course," Regi says, "but it's more than that. Green Man's garden isn't based off of permaculture, or at least not the work of Bill Mollison. It's the way that my father figured out to plant a field that wouldn't go fallow. My father is illiterate. He never learned agricultural theory or math, but he has great intuition and native intelligence and that is a power that we've forgotten we have."

"I'm not sure I understand."

"Let me put it this way," Regi says, "When William turned three years old he suffered from night terrors; he would run screaming through the house because he thought that dinosaurs were chasing him, or bees. He was very vulnerable; it broke my heart to look at his eyes full of fear.

"I didn't know how to help him and we didn't have the money for therapists, but I spent a long time in the library doing research. I learned a lot of things about dreams. I learned that when a person suffered from night terrors they were awake for a thousandth of a second, and then they were asleep for a thousandth of a second and that they vibrated back and forth between the two worlds so quickly that they merged together.

"I also learned about lucid dreaming, which is the ability to recognize when you are in a dream and to take control of it, and about dream priming which is the practice of setting yourself up to dream in different ways."

"So what did you do?" I ask.

"I gave William two gifts. One was a magic wand, and one was a butterfly net. Then we worked together on what to do with each. The magic wand could shoot magic to drive off the dinosaurs, and he could use the butterfly net to catch the bees." Regi smiles and shakes his head at the memory. "I think I didn't really know what I was doing," he says, "but that was a rudimentary kind of dream priming, and the thing is, it worked, and as I look back, a lot of the logic and common sense we used growing up came from similar experiences. We did not have the science, but what we did worked. For one we unlocked knowledge from the subconscious and dealt with many difficulties of daily life and traumas. William was able to take control of his night terrors with those tools, and eventually the dreams went away.

"I look around and I see a world that is suffering from night terrors. We're struggling to wake up to realities of climate change, poverty, racism. I hope that the system I've developed can work as a bit of dream priming, to open people's minds to the possibility of living in different ways. We don't have to buy into the narrative that the corporations are selling us. We know intuitively that the industrial model doesn't make sense—we can intuit that farmers should be growing food, right?

"We have to unlock the power of our intuition to move forward. We have to live lucidly and acknowledge what's around us all while looking to the future." Regi laughs, "That's not too much to ask is it?"

"It seems like a lot to ask from a farm system," I say. "What if it doesn't work?"

Regi thinks for a minute, "I told you about the fire in Northfield," he says, "when we lost our home."

"Yes?"

"You know what kept me up at night afterward? I wasn't worried that there would be another one. What scared me were the 'what if' questions. What if Nicktae and William had been home? She would have been asleep maybe four inches from the fire roaring outside the wall. William would have been in his room in the basement right next to the dogs.

"I had dreams for a long time where I'm standing there and the house is burning and I'm asking myself if I have the courage to go into the basement. If I even have the time to go back in before the house explodes…

"We can't dwell on the negative what-ifs, we have to focus on the what-ifs that bring possibilities for positive change. Like what if we can help change the way we farm to a system that regenerates instead of what we have today? What if by doing so we also restore the health to the soil, the water systems above and below ground, the health of people who eat the polluted food that dominates our markets today? Heck, that's already starting to happen in a serious way. All I'm doing is making that change accessible to more people. What if we were to realize as a society that the lies of the corporate propaganda that we have come to believe as truths are really the main inhibitor in the unlocking of a new system … of the investments that can realize our potential as farmers and consumers to restore real science as the basis for the agriculture systems of the future?

"Agriculture has changed the course of the world so many times already. If we all come together and give what we can, then we will change the world again and it will be a better world for everyone."

"Well," I say, "when you put it that way, it doesn't sound so hard."

"Oh, but it is," Regi says. "Don't get me wrong. It's a hard choice, but it's the only one we have."

Index

agrochemicals, 34
Aj Quen. 130
Allende, Isabells, xii
Andreassen, Per, xi-xii
Arbenz, Jacobo, xviii
Armas, Carlos Castillo, xviii
Arzu, Alvaro, xviii
Augsburg College, ix
Aztec Coffee, 155

Barcenas Villa Nueva, 66
bikers, 163-164
Borges, Jorge Louis, xii
Cafe de la Selva, 155, 158
Canadian volunteers, 112
cattle, 86
Central National School of
    Agriculture, ix
chemical salesperson, 89
chemical spraying, 77
Chickens, 1, 6, 17-19
chickens, 93
Chimaltenango, 132
CIA, xviii
cilantro, 43, 49
Civil war, Guatemalan, xvi, xviii-xix
coffee, fair trade, 155
convent, 80-81

corn 99
corn, 5-7,

DDT, 100
De Leon Carpio, Ramiro, xviii
Del Toro, Guillermo, xii
Dramatic Society, 48-56

Fair Trade Federation, ix
Finca Mirasol, 176
fire, 168-169, 184-185
fungus, 78

Garden of the World, 85-87
genetically-modified plants, 4
Germans, 104
GMOs, 4, 9-10
graduation, 90-92
Great Snake, 59-61
Guatemala City, 127
Guatemala City, 50, 79
Guatemala en Vivo (Guatemala
    Alive), 153
Guatemalan Revolutionary
    National Unity rebels, xviii
guitar, 118-120, 127

Haslett-Marroquin, Amy, xv-xvii, 115-122, 133-136, 146, 159, 168-170
henhouse, 93
house fire, 168-169
hydroponics, 3

Institute for Agriculture and Trade Policy (IATP), 154, 158

Jocotan, 103

Kukulkan, 71-73

La Fundación Rigoberta Menchu, 153
lightning, 83
Llorona, La, 10, 20

Main Street Project, 1-2, xvii, ix
marimba, 78, 79, 88, 127, 167
Menchú, Rigoberta, 154, 158
Mexican volunteers, 113
Mollison, Bill, 161, 183
Montessori, 145
Mother Blight, 27-30, 38-40
mushrooms, 72-73

National School of Agriculture, 45, 77
orphanage, 117

Pan-American Highway, 156
Peace Coffee, x, xi, 158
permaculture, 161, 184
pine cones, foraging, 46
Poptún, 50, 51, 54, 101
potato plants, 178-179

Poultry-Centered Regenerative Agriculture System, xvii
priesthood, 114

Quetzaltenango, 130

Resource Center of the Americas, 143
Ritchie, Mark, 167
Ritchie, Niel, 154, 167

Santa Apolonia, 111
scorpions, 21-22
Seven Stories, 183
Seward neighborhood, 147
Siguanaba, La, 10
snakes, 35-36
spider monkeys, 27-29
Tecun Uman, 48

United Fruit Company, xviii
United Nations, ix
Universidad de San Carlos' Agronomy School, 77
University of Minnesota, 132, 134
University of San Carlos, 127

Virgin of Candelaria, 44

weeding, 30
woodworking, 63-64
World Council of Indigenous Peoples, ix

Young People's Action Coalition (YPAC), 176